# Infectious Disease

T0171881

**Series Editor**

Vassil St. Georgiev
National Institute of Health Department Health & Human Services,
Bethesda, MD, USA

The Infectious Disease series provides the best resources in cutting-edge research and technology.

More information about this series at http://www.springer.com/series/7646

Jerome Goddard

# Infectious Diseases and Arthropods

Third Edition

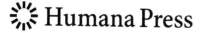 Humana Press

Jerome Goddard
Extension Professor of Medical Entomology
Mississippi State University
Mississippi State, MS, USA

Infectious Disease
ISBN 978-3-030-09346-4         ISBN 978-3-319-75874-9   (eBook)
https://doi.org/10.1007/978-3-319-75874-9

Printed on acid-free paper

This Humana Press imprint is published by the registered company Springer International Publishing AG
part of Springer Nature.
The registered company address is: Gewerbestrasse 11, 6330 Cham, Switzerland

*For Rosella, my inspiration*

# Preface

Infectious diseases aren't conquered. In fact, a good argument can be made that they are gaining ground in their long struggle with humankind. The ability of microbes to adapt to host immune responses and intense pressure from antibiotic use, combined with societal changes, has contributed to a resurgence of many infectious diseases. In addition, there are several "new" or emerging diseases, including Lyme disease, some forms of ehrlichiosis, Heartland and Powassan tick viruses, SARS, MERS, bird flu, Zika virus, Chikungunya, and Ebola hemorrhagic fever. In just the last 30 years or so we have seen the appearance of a virulent strain of avian influenza that attacks humans, a human variant of "mad cow" disease, and all manner of multiple drug-resistant bacteria such as *Staphylococcus aureus*. These new or emerging infectious diseases have raised considerable concern about the possibility of widespread and possibly devastating disease epidemics among human populations.

Many infectious diseases are vector-borne, i.e., carried from one host or place to another by an arthropod vector. It could be argued that at least some of the recent increase in vector-borne disease is the result of increased recognition and reporting. Specific disease and vector recognition is certainly made easier by newer technologies such as the polymerase chain reaction (PCR) and barcoding technologies. However, societal changes such as population increases, international travel, ecological and environmental changes, and especially suburbanization (building homes in tracts of forested lands) are contributing to an increase in the incidence of many of these vector-borne diseases.

In light of this vector-borne disease increase, information about these arthropod entities – their distributions, hosts, reservoirs, and vectors – is much needed. Thus, this third edition of *Infectious Diseases and Arthropods* is intended to provide physicians, as well as entomologists and other interested parties, with a reference on the biological and entomological aspects of infectious diseases. The primary approach has been to present readily accessible information on the major vector-borne diseases, with an emphasis on the relevant biology and ecology of each one. Since I am writing as an entomologist, the text obviously leans heavily to the organismal side of each disease, with, in some cases, less emphasis on clinical aspects. No effort has been made to present an in-depth review of each disease; instead, there is a middle-of-the road consensus of current thought on each subject. It is the author's hope that *Infectious Diseases and Arthropods*, Third Edition, will prove a useful adjunct to the larger clinical texts employed by infectious disease specialists, public health and travel medicine physicians, epidemiologists, and others with duties encompassing vector-borne diseases. Treatments are mentioned (but without specific dosages) for the various diseases, but are only intended as general guidelines. They are in no way intended to be the sole, specific treatment for any particular patient. Physicians should consult clinical texts or drug package inserts for the most current recommendations.

Mississippi State University, MS, USA                                        Jerome Goddard

# Acknowledgments

A substantial portion of this text was originally published (text, figures, and photos) as part of a series of medical entomology columns written for *Infections in Medicine*. For inclusion here, these were updated with new and current medical and scientific information. Specific credits for the columns are listed below. This material is kindly reprinted by permission of CMP Healthcare Media, Darien, CT, publishers of *Infections in Medicine*. Also, part of Chap. 1 was originally published as an article entitled "Arthropods and Medicine" in *Agromedicine*, and is reprinted here by permission from Hayworth Press, Inc., Binghamton, NY.

The following people read portions of the manuscript (either first, second, or third edition), supplied portions of text, and/or offered helpful advice: Michelle Allerdice (Centers for Disease Control), Rosella M. Goddard, Dr. Hans Klompen (Ohio State University), Dr. Bill Lushbaugh (formerly at the University of Mississippi School of Medicine), Dr. Chad P. McHugh (formerly at Brooks Air Force Base, San Antonio, TX), and Dr. Andrea Varela-Stokes (Mississippi State University). The following people allowed me to use their photographs and/or arthropod specimens for pictures contained in this book: Tom and Pat Baker (Oakdale, CT), Dr. Lorenza Beati (U.S. National Tick Collection), Dr. Kristine Edwards (Mississippi State University), Mallory Carter, Joseph Goddard (Alabama Department of Wildlife and Fisheries), Dr. Blake Layton (Mississippi State University), Joe MacGown (Mississippi State University), and Wendy C. Varnado (Mississippi Department of Health).

Rosella M. Goddard created the Third Edition worldwide distribution maps for the various arthropod vectors and diseases. I am especially grateful to her for that.

The following *Infections in Medicine Bug Vectors* columns were included in the First Edition of this book and serve as the basis for all subsequent editions:

- Arthropods and Medicine, 1996; 13: 543, 544, 557, © 1996, CMP Healthcare Media LLC
- Eastern Equine Encephalitis, 1996; 13: 670–672, © 1996, CMP Healthcare Media LLC

- St. Louis Encephalitis, 1996; 13: 747, 751, 806, © 1996, CMP Healthcare Media LLC
- Dengue Fever, 1996; 13: 933, 934, 984, © 1996, CMP Healthcare Media LLC
- Rocky Mountain Spotted Fever, 1997; 14: 18–20, © 1997, CMP Healthcare Media LLC
- Ehrlichiosis, 1997; 14: 224, 229, 230, © 1997, CMP Healthcare Media LLC
- Mosquitoes and HIV, 1997; 14: 353, 354, © 1997, CMP Healthcare Media LLC
- Malaria, 1997; 14: 520–522, 547, © 1997, CMP Healthcare Media LLC
- Lyme Disease, 1997; 14: 698–700, 702, © 1997, CMP Healthcare Media LLC
- Tick-Borne Viruses, 1997; 14: 859–861, © 1997, CMP Healthcare Media LLC
- Tick Paralysis, 1998; 15: 28–31, © 1998, CMP Healthcare Media LLC
- Imaginary Insect or Mite Infestations, © 1998; 15: 168–170, © 1998, CMP Healthcare Media LLC
- Tularemia, 1998; 15: 306–308, © 1998, CMP Healthcare Media LLC
- Murine Typhus, 1998; 15: 438–440, © 1998, CMP Healthcare Media LLC
- Lymphatic Filariasis, 1998; 15: 607–609, © 1998, CMP Healthcare Media LLC
- Yellow Fever, 1998; 15: 761, 762, 765, © 1998, CMP Healthcare Media LLC
- Plague, 1999; 16: 21–23, © 1999, CMP Healthcare Media LLC
- Chagas' Disease, 1999; 16: 23–26, © 1999, CMP Healthcare Media LLC
- Ticks and Human Babesiosis, 1999; 16: 319–320, © 1999, CMP Healthcare Media LLC
- Leishmaniasis, 1999; 16:566–569, © 1999, CMP Healthcare Media LLC
- Tick-Borne Relapsing Fever, 1999; 16: 632–634, © 1999, CMP Healthcare Media LLC
- Chigger Bites and Scrub Typhus, 2000; 17: 236–239, © 2000, CMP Healthcare Media LLC
- LaCrosse Encepalitis, 2000; 17: 407–410, © 2000, CMP Healthcare Media LLC
- Human Lice and Disease, 2000; 17: 660–664, © 2000, CMP Healthcare Media LLC
- Myiasis Confused with Boils, 2001; 18: 17–19, © 2001, CMP Healthcare Media LLC
- Black Flies and Onchocerciasis, 2001; 18: 293–296, © 2001, CMP Healthcare Media LLC
- Health Problems from Kissing Bugs, 2003; 20: 335–338, © 2003, CMP Healthcare Media LLC
- Bed Bugs Bounce Back – But Do They Transmit Disease? 2003; 20: 473–474, © 2003, CMP Healthcare Media LLC
- American Boutonneuse Fever, 2004; 21: 207–210, © 2004, CMP Healthcare Media LLC
- Fire Ant Attacks on Humans, 2004; 21: 587–589, © 2004, CMP Healthcare Media LLC
- Brown Recluse Spider Bites, 2005; 22: 205–210, © 2005, CMP Healthcare Media LLC
- Fleas and Human Disease, 2005; 22: 402–405, © 2005, CMP Health

# Contents

# About the Author

**Jerome Goddard** is a medical entomologist currently located at Mississippi State University who is known for research on a number of medically important arthropods, most notably ticks, mosquitoes, and bed bugs. Prior to coming to Mississippi State, Dr. Goddard was an Air Force medical entomologist for 3 years and then state medical entomologist for the Mississippi Department of Health for 20 years. After Hurricane Katrina (2005), he was the health department official responsible for the mosquito and vector control program along the Mississippi Gulf Coast. He is the author of over 200 scientific articles and a medical textbook used by physicians, which won "Highly Commended" in 2003 in the British Medical Association's Best Medical Book of the Year competition. Over the last two decades, he has served as an educational resource concerning medically important arthropods to a U.S. Congressional Committee, in various newspapers and magazines such as Reader's Digest, and on television programs such as The Learning Channel ("Living with Bugs") and the Colbert Report.

# Part I
# Arthropods and Human Health

# Chapter 1
# Arthropods and Health

## 1.1 Classification of Arthropods

The phylum Arthropoda includes insects, spiders, mites, ticks, scorpions, centipedes, millipedes, crabs, shrimp, lobsters, sow bugs (roly-polies), and other related organisms. Arthropods are characterized by segmented bodies; paired, jointed appendages (e.g., legs and antennae); an exoskeleton; and bilateral symmetry (Fig. 1.1) [1]. Arthropods display an amazing diversity and abundance. They make up more than 85% of all known animal species [2]. Arthropods are found on every continent, and a square meter of vegetation is literally teeming with them. For brevity, four classes of arthropods will be discussed in this chapter—insects, arachnids, centipedes, and millipedes. Table 1.1 discusses some key characteristics of these major arthropod groups.

### 1.1.1 Insects

Like other arthropods, insects possess a segmented body and jointed appendages. Beyond that, however, there is much variation: long legs or short legs; four wings, two wings, or no wings; biting mouthparts or sucking mouthparts; and soft bodies, hard bodies, etc. Despite the diversity, adult specimens can be recognized as insects by having three pairs of walking legs and three body regions: a head, a thorax (bearing legs and wings if present), and an abdomen. No other arthropods have wings. Although most adult insects have wings, several medically important species are wingless (e.g., lice, fleas, bed bugs).

Insects have different forms of development. In those with gradual metamorphosis (grasshoppers, lice, true bugs), the immatures are called nymphs and are structurally similar to the adults, increase in size at each molt, and develop wings (if present) during later molts (Fig. 1.2). In groups with complete metamorphosis

© Springer International Publishing AG, part of Springer Nature 2018
J. Goddard, *Infectious Diseases and Arthropods*, Infectious Disease,
https://doi.org/10.1007/978-3-319-75874-9_1

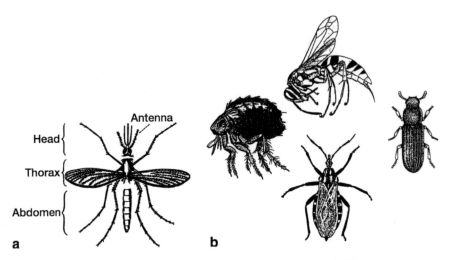

**Fig. 1.1** (**a**) Generalized insect drawing with parts labeled; (**b**) several different insect types. (From US Department HEW [CDC] pictorial keys)

(e.g., beetles, flies, bees and wasps, moths and butterflies, and fleas), the immature stages are called larvae and pupae and look nothing like the adult (Fig. 1.3). Often, larvae are wormlike and are frequently called "worms" by lay people (Fig. 1.4). The three body regions are never as distinct as they are in adults, but generally three pairs of short walking legs are evident. Fly larvae (maggots) lack walking legs, and although some such as mosquitoes have three body regions, others (e.g., larvae of houseflies and blowflies) do not have distinct body regions. Caterpillars and similar larvae often appear to have legs on some abdominal segments. Close examination of these abdominal "legs" (prolegs) reveals that they are unsegmented fleshy projections, with or without a series of small hooks (crochets) on the plantar surface, and structurally quite unlike segmented walking legs on the first three body segments behind the head.

### 1.1.2 Spiders

Spiders have two body regions—an anterior cephalothorax and a posterior abdomen connected by a waist-like pedicle (Fig. 1.5). The anterior portion consists of the head with various numbers of simple eyes on the anterior dorsal surface and the thorax with four pairs of walking legs. The mouthparts, called chelicerae, are hollow, sclerotized, and fang-like and are used to inject venom into prey. Located between the chelicerae and the first pair of walking legs are a pair of short leglike structures called pedipalpi, which are used to hold and manipulate prey. Pedipalpi may be modified into copulatory organs in males. The spider abdomen is usually unsegmented and displays spinnerets for web production at the posterior end. Immatures look the same as adults, except smaller.

**Table 1.1**  Key characteristics of some arthropod groups

| Arthropod group | Class | Characteristics | Remarks |
|---|---|---|---|
| Insects | Insecta or Hexapoda | Six legs | Mostly nonharmful (even helpful) to humans; some species bite/sting or transmit disease organisms |
| | | Three body regions most with wings | |
| Spiders | Arachnida | Eight legs | Most able to bite, but with little or no consequence |
| | | Two body regions: Cephalothorax, abdomen | Brown recluse, widow spiders, hobo spider may be dangerous in the United States |
| Mites and ticks | Arachnida | Eight legs (as adults) | Ticks are essentially "large mites" |
| | | One globose or disk-shaped body region | Ticks transmit many different disease agents to humans and may cause paralysis while feeding |
| | | No true heads, mouthparts only | |
| Scorpions | Arachnida | Eight legs | Only one dangerous species in the United States occurring in Arizona and New Mexico |
| Centipedes | Chilopoda | One pair of legs per body segment | Called "hundred-leggers" |
| | | Often dorsoventrally flattened | Painful bites but mostly harmless |
| Millipedes | Diplopoda | Two pairs of legs per body segment | Called "thousand-leggers" |
| | | Often cylindrical | Defensive fluids may cause burns or stains on the skin |

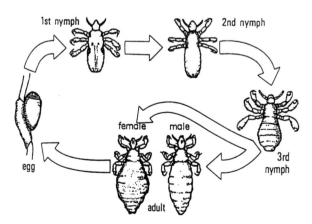

**Fig. 1.2** Head lice life cycle, example of a gradual metamorphosis. (From US, DHHS, CDC, home study course 83–3297)

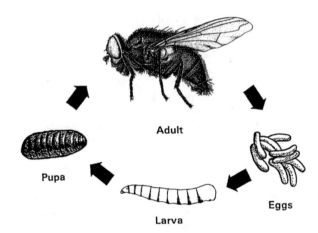

**Fig. 1.3** House fly life cycle—example of complete metamorphosis. (From USDA, ARS, Agri. Hndbk. No. 655, Feb. 1991)

One note must be added about "daddy longlegs," since most people erroneously call them spiders. Harvestmen, or daddy longlegs (order Opiliones), have many characteristics in common with true spiders; however, they differ in that the abdomen is segmented and broadly joined to the cephalothorax (not petiolate). Most species have extremely long, slender legs. Contrary to folklore, they are not venomous.

### 1.1.3  Mites and Ticks

These small arachnids appear to have only one body region (cephalothorax and abdomen fused), the overall appearance being globose or disk-shaped (Fig. 1.6). This general appearance quickly separates them from other arthropods. The body may be segmented or unsegmented with eight walking legs present in adults. Larvae, the first-stage immatures, have only six (rarely fewer) legs, but their fused cephalothorax and abdomen readily separates them from insects. They attain the fourth pair of legs at the first molt and thereafter are called nymphs until they become adults. As for spiders, immature ticks and mites are generally similar in appearance to adults. In general, ticks are considerably larger than mites. In fact, ticks are just large mites. Adult ticks are generally about the size of a pea; mites are about the size of a grain of sand (often smaller).

### 1.1.4  Scorpions

Scorpions are dorsoventrally flattened creatures with an anterior broad, flat area and a posterior segmented "tail" with a terminal sting (Fig. 1.7). Although these outward divisions do not correspond with actual lines of tagmatization, they do provide an

**Fig. 1.4** Various types of insect larvae. (From US, DHEW, PHS, CDC, pictorial keys)

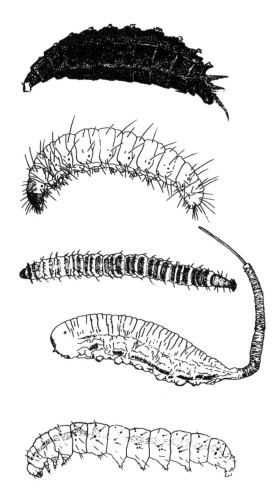

appearance sufficient to distinguish these arthropods from most others. Like spiders, they have four pairs of legs, the mouthparts are chelicerae, and the first elongate appendages are pedipalpi. Scorpion pedipalpi are modified into pinchers to capture prey. Immatures are similar to adults in general body form.

### 1.1.5 Centipedes and Millipedes

Centipedes and millipedes bear little resemblance to the other arthropods previously discussed. They have hardened wormlike bodies with distinct heads and multiple pairs of walking legs (Figs. 1.8 and 1.9). Centipedes are swift-moving, predatory organisms with one pair of long legs on each body segment behind the head. Millipedes, on the other hand, are slow-moving omnivores or scavengers that have two pairs of short legs on each body segment (after the first three segments, which only have one pair each). Immatures are similar to the adults.

**Fig. 1.5** Various spiders:
top, tarantula; middle,
brown recluse; bottom,
black widow dorsal view
(not drawn to scale). (US,
DHEW, PHS, CDC,
pictorial keys)

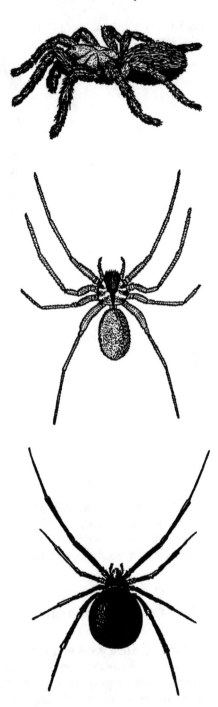

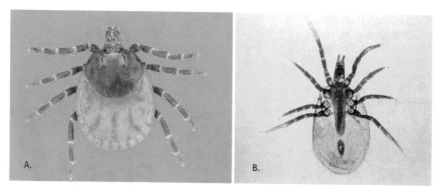

**Fig. 1.6** Tick (**a**) and mite (**b**); not drawn to scale. (Tick photo courtesy Dr. Blake Layton, Mississippi State University and mite photo courtesy the CDC)

**Fig. 1.7** Typical scorpion. (Photo copyright 2008 by Jerome Goddard Ph.D.)

## 1.2   Identification Methods for Arthropods

### 1.2.1   Morphological Identification

The historical method for identifying arthropods is performed by examining various structures (the morphology) on the specimens and making a classification based on size, shape, or number of those characters. This process often involves using published diagnostic keys wherein one follows a flowchart to arrive at the proper identification. Good-written keys include numerous high-quality photos or line drawings

**Fig. 1.8** Centipedes and millipedes: (**a**) common house centipede, (**b**) giant centipede, (**c**) common millipede. (From US, DHEW, PHS, CDC, pictorial keys)

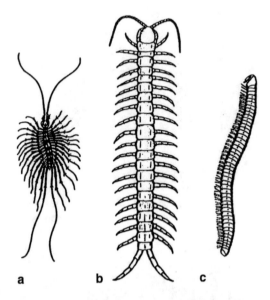

a            b            c

**Fig. 1.9** Centipede. (Photo copyright 2016 by Jerome Goddard, Ph.D.)

of the structures mentioned (sometimes line drawing is actually better at showing detail than photographs). A written diagnostic key may look something like this:

1. Second antennal segment without a laterodorsal longitudinal seam . . . Acalypteratae
1' Second antennal segment with a complete laterodorsal longitudinal seam . . . go to 2
2. Hypopleuron (Meron) usually without hairs or bristles . . . Anthomyiaria
2' Hypopleuron (Meron) usually with hairs or bristles in one or more rows . . . go to 3

On the other hand, a picture key may look something like the one presented in Fig. 1.10.

Identification keys may continue on like this for hundreds of choices or couplets comprising many pages. As is immediately obvious, keys can use complex

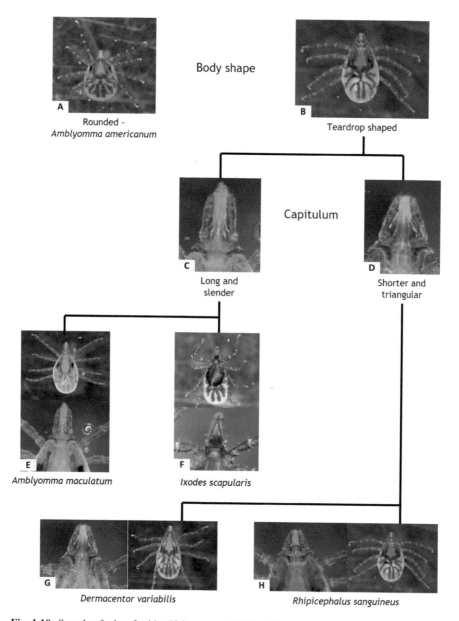

**Fig. 1.10** Sample of a key for identifying nymphal ticks. (Photo courtesy Drs. Trisha Dubie and Bruce Noden, Oklahoma State University, used with permission)

terminology that lay people are not familiar with, making identification difficult. Well-written keys avoid jargon specific to any one group of arthropods and try to stay clear of couplets with vague phrases such as "such and such structure is longer." If the person making the identification cannot see a specimen with the second option, they cannot make a decision. How long is longer? Therefore, morphological characteristics in a key should be quantitative whenever possible. Lastly, making a morphological identification does not harm the specimen, so voucher specimens can be deposited in a museum and kept for future examination if there is ever a question.

### 1.2.2   Molecular Identification

Molecular identification of arthropods involves looking at their genetic material by grinding up the arthropod in question (or pieces thereof), running a polymerase chain reaction (PCR) procedure to obtain DNA of portions of one or more key genes of that specimen and then comparing those gene sequences with known (published) sequences found in the National Center for Biotechnology Information, National Library of Medicine, and "GenBank." Searching GenBank can reveal what the sequences most closely resemble. For example, if you did a PCR on an unidentified tick specimen using published primers for identifying ticks and then submitted your sequence to GenBank, you would get a result showing the closest matches. Your result may say "100% match with *Amblyomma americanum*." If so, then that is the precise identification. If your result says "closest match is *Amblyomma parvum*, 86%," then you can assume something went wrong with the analysis, or perhaps you have a specimen with no sequences in GenBank, or maybe you have found a new species. To improve the accuracy of molecular identifications, more than one gene should be used; the more the better.

In the last decade, there has been an effort to barcode all animal life using the cytochrome c oxidase gene. Use of these DNA barcodes has supposedly been validated [3], although not everyone agrees [4]. According to Will [4], people seeking an immediate panacea of molecular identification of species will encounter all sorts of constraints and inconsistencies in their work, most importantly, judgements about species boundaries. This problem is discussed in a paper by Sperling [5], wherein he discusses what DNA sequence or allozyme divergence number is the cutoff above which populations can be considered separate species. Based on his data from *Papilio* butterflies, Sperling says it is unreasonable to expect any kind of simple relationship between percent sequence divergence and the maintenance of genetic integrity [5].

### 1.2.3   Pros and Cons of Morphological Versus Molecular Identification

Some questions cannot be answered with a molecular analysis of an arthropod. The best example is developmental stage; in morphological identification, one can almost instantly determine the life stage of an arthropod. This determination is more difficult using molecular methods. Certain arthropods contain different concentrations of mRNA at different time points in their development, in which case these genes can be amplified by PCR and analyzed for concentration; however, this is often not the case. If identification to developmental stage is required, morphological analysis is the best method to use. If developmental stage is not a point of concern and identification of species is the main concern, then maybe the more robust method is molecular analysis. However, genetic material is not always available, especially from historic collections. Additionally, molecular analysis is expensive and time-consuming; a laboratory needs access to DNA extraction kits, reagents for amplification, a thermal cycler to perform the amplification, access to sequencing capacity or electrophoresis, and the software required to analyze completed sequences. In addition to the expense, a complex understanding of the organism's genome is necessary to successfully use molecular analysis. For example, many hard ticks are closely related genetically, and some loci differentiate between species by only a few base pairs. If the targeted region of the genome is not informative between related species, this differentiation can be convoluted and inconclusive.

As mentioned, while molecular analysis can potentially provide clear distinction between species, these analyses are limited to the availability of genetic data. GenBank has a huge cache of genetic data with which to compare results, but there is no review process to submit genetic data to GenBank. This has allowed for at least a proportion of the data in GenBank to be not accurate, and how to determine what is and is not accurate is not simple. Studies have shown that up to 30% of the data publicly available in GenBank contains errors. If a BLAST search reveals numerous identical results from multiple sources, then chances are good the identification is correct; however, if only a few results are produced, confidence in the identification should be low. Another major point to consider when using molecular sequencing for species identification is that origin of all genetic data in GenBank is, in fact, morphological. Because GenBank is an open forum for submission, whoever submits the first sequence for an arthropod species designates it based on his own morphological identification. If the arthropod was misidentified, the data associated with it is incorrect. This can become a serious issue, as corrections may rarely be made if an original submission to GenBank was misidentified. No submissions to GenBank ever require publication, so submissions are thereby immune from peer review. Therefore, a misidentification could potentially persist unnoticed or uncorrected for decades.

Both morphological and molecular identification techniques have benefits and disadvantages, but a prudent decision to use one or the other must be made based on the available equipment, knowledge, and resources. In addition, the purpose of the identification must be considered; if the determination of developmental stage is required, or if physical characteristics or general fitness are a point of concern, morphological analysis will provide the best information. However, if a robust identification or secondary verification of a species is needed, molecular analysis will work well. Ideal situations employ both methods together and provide the most data; in this way, one can consider both physical characteristics and genetic information in tandem to make the most accurate determination. In either case, if identification seems to be an anomaly, further analysis is necessary.

## 1.3  Medical Importance of Arthropods

Arthropods may affect human health directly or indirectly. Directly, humans are affected by bites, stings, myiasis, blistering/staining, and other mechanisms such as insects getting inside the ears [6]; indirectly, they are affected through allergies and disease transmission (*see* Chap. 2 for a detailed list). However, one must be careful not to consider all arthropods detrimental or dangerous. Only a small percentage are medically important. Most arthropods are benign as far as their effects on humans are concerned and are extremely important components in ecological communities, serving as food sources for other creatures and as pollinators.

### 1.3.1  Historical Aspects of Medically Important Arthropods

Humans have undoubtedly been bothered by arthropods since prehistoric times. Recorded instances of arthropod-borne diseases and infestations go back to the Old Testament, in which accounts of plagues on the Egyptians are described, many of which were apparently caused by insects. Also, about 2500 BC a Sumerian doctor inscribed on a clay tablet a prescription for the use of sulfur in the treatment of itch (sulfur is now known to kill itch and chigger mites) [7]. First-century BC hair combs containing remains of lice and their eggs have been unearthed [8]. Some Peruvian pottery from circa 600 AD shows natives examining their feet—and their feet display what appear to be holes where chigoe fleas (burrowing fleas) have been removed [9]. Other pottery found near the Mimbres River, New Mexico, dated to circa 1200 AD, clearly depicts a swarm of mosquitoes poised for attack. Modern medical entomology had its beginning in the late 1800s. In the space of about 20 years, several fundamental discoveries were made linking arthropods with the causal agents of disease. This opened a whole new vista, the so-called vector-borne diseases, but lest we develop chronological snobbery, thinking that our ancestors were "less enlightened" or somehow unintelligent, considering the fact that in 1577, Mercurialis believed that flies might carry the agent of plague from ill to healthy persons. In addition, in 1764, the physician Cosme Bueno described the conditions of cutaneous leishmaniasis and Carrion's disease in Peru and attributed them both to the bite of a small insect called uta [10]. The word "uta" is still used sometimes in the Peruvian highlands for sand flies, the vector of leishmaniasis. Sand flies are small and inconspicuous, and it is amazing that anyone could make that connection. Thousands of years ago, before the routine collection and recording of information, there may have been other insights into transmission of disease pathogens by arthropods.

Arthropods themselves, as well as the disease agents they transmit, have greatly influenced human civilization. Sometimes the influence has been notable or recorded, such as when plague epidemics swept through Europe or louse-borne typhus decimated armies. A more recent disintegration of society is described by Crosby in her book about the yellow fever epidemic in Memphis, TN (Fig. 1.11) [11]. However, in many other instances, the influence of arthropods has not been easily recognized.

**Fig. 1.11** Tombstone of person who died of yellow fever in Mobile, AL, 1853. Interestingly, the hole in the top of this tombstone is a breeding site for the mosquito that transmits yellow fever. (Photo copyright 2017 by Jerome Goddard, Ph.D.)

Great expanses of seacoast areas (e.g., Florida) or inland swamp areas were left undeveloped because of fierce and unbearable mosquito populations. These areas were only populated after the advent of effective area-wide mosquito control. In a similar fashion, a large part of Africa was left untouched by humans for centuries because of the risks posed by African trypanosomiasis (sleeping sickness) and falciparum malaria.

**Early Beginnings of Public Health Pest Control**

The earliest written history reveals that health and hygiene issues were assigned to priests and other religious leaders. There was a position in the Egyptian government over 5000 years ago called the *vizier*, a priestly minister of state, who had the duty of inspecting the water supply every few days. The role of biblical Hebrew priests was similar, albeit more sophisticated and elaborate. Although its origin may have been derived somewhat from Egyptian and Mesopotamian medical knowledge, the Mosaic Code developed by the Hebrews is the basis of all modern sanitation and pest control to this day. This code was the most successful means of achieving health devised by humans for nearly 3000 years [12] and included unique insights such as establishing quarantines, washing of hands with running water, avoiding contact with the dead, and eating meat only from healthy animals killed and processed under close scrutiny of the rabbis. Later, science, medicine, philosophy, and the arts flourished in ancient Greece for almost 500 years. Asclepius was the patron saint of Greek medicine, and his daughter, Hygeia, was the saint of preventive medicine. During the Roman Empire, Galen's medical writings became the authoritative source for all late-medieval medicine. The Romans made great strides in sanitation, civil engineering, and hygiene, with construction of aqueducts and sewage disposal systems. They were also among the first to develop indoor sanitation systems and baths. The Roman position of *aediles* (about 500 BC) was a non-physician administrator of sanitation, which included the task of caring for the safety and well-being of the City of Rome, and specifically, watching over public buildings, streets, aqueducts, and sewers, and maintaining various fire prevention strategies. After the Roman Empire during the Dark Ages, science-based policies gradually disappeared from public life and, along with them, much of the sanitation and health-related arts. There are reports that monks and other church leaders rarely bathed, considering cleanliness as an abhorrence, and according to Snetsinger [12], they were vermin-ridden. Plague, leprosy, smallpox, and tuberculosis emerged among the European population from about 800 AD to 1400 AD. However, public health and sanitation ideas survived those difficult times and began to reemerge (Fig. 1.12). For example, in England in 1297, there were regulations requiring every person to keep the front of his house or apartment clean, and in 1357, a royal order was issued by King Edward III requiring the mayor of London to prevent pollution of the Thames River [13]. King John of France in 1350 established a governmental entity called "sanitary police" [13]. Gradually, concepts of cleanliness and hygiene returned as important factors in society. In 1491, Johann Pruss promoted regular bathing and changes of clothing for deterrent of lice, and Ignatius Loyola, the famous Catholic priest, established rules of nutrition, exercise, and cleanliness for his followers. In 1518, a royal proclamation was issued by Henry VIII setting out ways to help control plague, such as marking of affected houses [14]. In the late 1500s, Queen Elizabeth I instituted housing regulations to relieve overcrowding, pure food laws to protect against adulteration of foodstuffs, and further regulations for epidemic control. During this time, arthropod and rodent pests became recognized as being related to filth, waste, and harborage. Slowly but surely, efforts were made to properly manage

**Fig. 1.12** From 1200 to 1500 AD, ideas about public health and sanitation began to reemerge in Europe. (Photo depicting a woman from St. Andrew's Church, Blickling, England, 1485. Brass rubbing made by Cathy Windham Beckett in 1975; photo copyright 2017 by Jerome Goddard, Ph.D., used with permission)

**Fig. 1.13** Outhouse for
waste disposal

human wastes without just dumping them into local rivers (Fig. 1.13). The
Renaissance opened the door for even more improvements in sanitation and hygiene
in the human population which have continued to this day.

## 1.3.2   Direct Effects on Health

Arthropod stings and bites cause significant pain and suffering each year [15–18].
In fact, one study reported over 620,000 insect bites/stings treated in emergency
departments each year [19]. Most stings result when social insects, such as bees,
ants, and wasps, defensively attack persons coming near or disturbing their nests
(Fig. 1.14). Venom is injected on stinging. Thus, the term envenomization (or envenomation) is an accurate descriptor. Venoms vary in chemical content from species to
species but basically contain highly complex mixtures of pharmacologically active
agents, biologically active agents, or both (e.g., histamine, serotonin, dopamine,
melittin, apamin, kinins, and enzymes, such as phospholipase A) [20]. Imported fire
ants are somewhat different, having an alkaloidal venom. Scorpion venom characteristically contains multiple low-molecular-weight basic proteins (the neurotox-
ins), mucus (5–10%), salts, and various organic compounds, such as oligopeptides,
nucleotides, and amino acids [21].

   Bites may result in significant lesions as well, but not because of injected venom
(except for spiders). Bite lesions are generally a result of immune reactions to
salivary secretions injected during the biting process (Figs. 1.15 and 1.16) [22].

**Fig. 1.14** Wasp sting.
(Photograph copyright
2012 by Jerome Goddard,
Ph.D., and courtesy of
Audrey Sheridian)

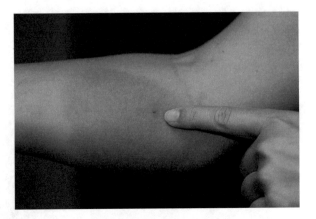

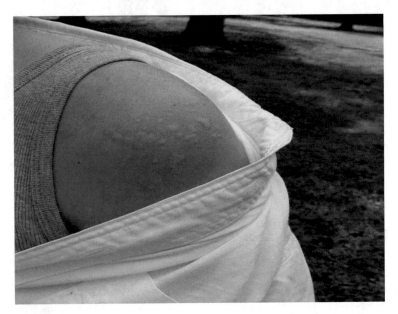

**Fig. 1.15** Mosquito bite lesions showing inflammatory response. (Photo copyright 2005 by Jerome Goddard, Ph.D.)

Arthropods inject saliva to lubricate the mouthparts on insertion, suppress host immune responses, increase blood flow to the bite site, inhibit coagulation of blood, aid in digestion, or a combination of factors. Humans may become hypersensitive to salivary secretions from groups of arthropods (e.g., mosquitoes or bed bugs) after repeated exposure. Spiders inject a venom, ordinarily used for killing and digesting the soft tissues of prey, which may cause neurotoxic effects (e.g., black or brown widow spider venom) or necrotic effects (e.g., fiddle back or hobo spider venom).

Myiasis occurs when fly larvae (maggots) infest the tissues of people or animals. It is mostly accidental or opportunistic, but in a few tropical species, the myiasis is purposeful, or obligate, with the larvae requiring time inside host tissues

**Fig. 1.16** Large indurations from mosquito bites. (Photo courtesy Dr. Wendy C. Varnado, Mississippi Department of Health, used with permission)

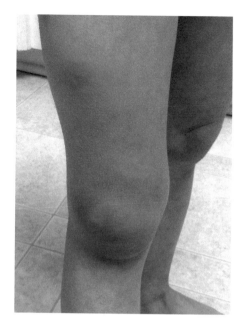

for development. Except for some cases of obligate cases (e.g., screwworm fly or bot fly), myiasis is generally not life-threatening. Interestingly, some fly larvae have been used in the past and currently are used on a limited basis by the medical profession to debride wounds [23]. These maggots only eat dead tissue and produce antibiotic substances, which reduce infection.

Some beetles, called blister beetles, possess the chemical cantharidin in their body fluids which can produce large fluid-filled blisters when the beetles come into contact with the human skin (Fig. 1.17) [17]. Fluids are secreted when the beetles are touched or handled. However, most blistering occurs when people hit or smash the insects on their bodies. Some millipede species also can cause stains or burns on human skin via defensive body fluids [10].

## 1.3.3   Indirect Effects on Health

### 1.3.3.1   Disease Transmission

Disease transmission is the primary indirect effect of arthropods on human health. The bite itself causes no health problem—it is the etiologic agent transferred during the event. Depending on incubation period, development of disease may not occur for days or months. Disease transmission by arthropods involves many interacting factors, such as presence and behavior of animal reservoir hosts, competence of arthropod vectors, and host/pathogen interactions (*see* Chap. 2). An understanding of how disease pathogens are acquired and transmitted by arthropods is crucial to preventing and/or managing vector-borne diseases.

**Fig. 1.17**   Blister beetle. (Photo copyright 2005 by Jerome Goddard, Ph.D.)

Often unnoticed by practicing physicians in the temperate zone, arthropod-borne diseases account for a huge portion of the spectrum of human maladies worldwide, and in some areas the problem is growing. Millions of people are at risk of African trypanosomiasis (sleeping sickness) and American trypanosomiasis (Chagas' disease) [24, 25]. Dengue fever, transmitted by mosquitoes, is epidemic throughout much of the Caribbean, Mexico, and Central and South America [26]. There are an estimated 100 million cases of dengue fever each year and about 500,000 cases of dengue hemorrhagic fever [27, 28]. Some countries have reported a 700-fold increase over the past 30 years [28]. Although recently decreasing in case numbers, malaria continues to be a huge burden on society, especially in Africa and Southeast Asia. There are several hundred million new cases each year with about 500,000 deaths, mostly young children [29]. To make matters worse, for most of these diseases, not only are the mosquito vectors becoming resistant to the insecticides used for their control, but the parasites/pathogens are becoming increasingly resistant to the antimicrobial drugs used to destroy them. Other vector-borne diseases appear to be emerging [30, 31]. Lyme disease, unknown until the late 1970s, now accounts for approximately 40,000 cases of tick-borne disease each year [32]. Thousands of cases of human ehrlichiosis have occurred since the first case was recognized in 1986 [32, 33].

### 1.3.3.2   Arthropod Allergy

Numerous arthropods can cause allergic reactions in persons by their stings, including various wasps, bees, ants, scorpions, and even caterpillars. However, the ones most commonly involved are paper wasps, yellowjackets, honeybees, and fire ants.

In addition to stings, bites from some arthropods may produce allergic reactions, including anaphylaxis and other systemic effects. However, systemic hypersensitivity reactions to arthropod bites are much less common (almost rare) than those resulting from stings. The groups most often involved in producing systemic effects by their bites are the kissing bugs (genus *Triatoma*), tsetse flies, black flies, horse flies, and deer flies (Fig. 1.18), although bed bugs have recently been reported as a cause of systemic reactions [34–37]. Mosquitoes, to a lesser extent, are involved, with several reports in the literature of large local reactions, urticaria, angioedema, headache, dizziness, lethargy, and even asthma [38]. Tick bites may sometimes cause extensive swelling and rash. Ticks reported to do so are the hard ticks, *Ixodes holocyclus* and *Amblyomma triguttatum*, and the soft tick, *Ornithodoros gurneyi*. Arthropod saliva from biting insects contains anticoagulants, enzymes, agglutinins, and mucopolysaccharides. Presumably, these components of saliva serve as sensitizing allergens.

**Red Meat Allergy from Ticks**
Recently, researchers discovered a link between bites of ticks and subsequent development of allergies to red meat, the so-called red meat allergy [39, 40]. This happens when ticks (primarily the lone star tick, *Amblyomma americanum*) feed on mammals and acquire the carbohydrate galactose-$\alpha$-1,3-galactose or simply "alpha-gal" in the blood meal. Then, after molting to the next stage, if these ticks feed on a human, there is exposure to alpha-gal via tick saliva. If allergic, the person develops sensitivity to alpha-gal and then days or weeks later, when eating red meat, may develop urticaria, angioedema, or anaphylaxis 3–6 hours after red meat ingestion [41]. Meats known to express alpha-gal and fall into the "red meat" category include meat of cow, pig, lamb, squirrel, rabbit, horse, goat, deer, kangaroo, seal, and whale. Persons sensitive to alpha-gal should be able to eat chicken, turkey, and fish [41].

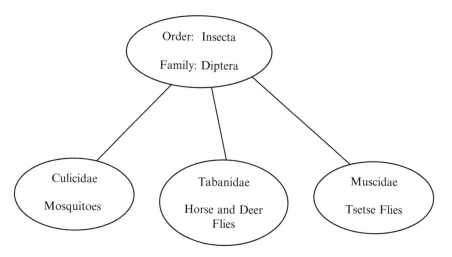

**Fig. 1.18** Classification of some flies known to cause allergic reactions by their bites. (Note: tsetse flies are now often placed in their own family – Glossinidae)

**Allergy/Irritation Caused by Consuming or Inhaling Arthropod Parts**

Several insect or mite species (or their body parts) may cause irritation and/or allergic reactions when inhaled and, less commonly, when ingested. House dust mites, *Dermatophagoides farinae* (and *D. pteronyssinus*), cockroaches, mayflies and caddisflies, and some nonbiting chironomid midges are major inhalant offenders. As these arthropods die, their decaying cast skins become part of the environmental dust [42]. In addition, insect emanations such as scales, antennae, feces, and saliva are suspected as being sources of sensitizing antigens. Compounding the problem, the average child spends 95% of his or her time indoors, providing plenty of time for sensitization. As for the digestive route, cockroach vomit, feces, and pieces of body parts or shed skins contaminating food are most often the cause of insect allergy via ingestion.

Until the mid-1960s, physicians simply diagnosed certain people as being allergic to house dust. Subsequently, Dutch researchers made the first link between house dust allergy and house dust mites [43, 44]. The mites commonly infest homes throughout much of the world and feed on shed human skin scales, mold, pollen, feathers, and animal dander. They are barely visible to the naked eye and live most commonly in mattresses and other furniture where people spend a lot of time. The mites are not poisonous and do not bite or sting, but they contain powerful allergens in their excreta, exoskeleton, and scales. For the hypersensitive individual living in an infested home, this can mean perennial rhinitis, urticaria, eczema, and asthma, often severe. In fact, Htut and Vickers [45] say that house dust mites are the major cause of asthma in the United Kingdom. House dust mites can also be triggers for atopic dermatitis [46].

Evidence indicates that early and prolonged exposure to inhaled allergens (such as dust mites and cockroaches) plays an important role in the development of both bronchial hyperreactivity and acute attacks of asthma. Accordingly, bronchial provocation with house dust mite or cockroach allergen can increase nonspecific reactivity for days or weeks. So, the root cause of asthma onset is sometimes the result of exposure to house dust mites or cockroaches. Asthma-related health problems are most severe among children in inner-city areas. It has been hypothesized that cockroach-infested housing is at least partly to blame [42]. In one study of 476 asthmatic inner-city children, 50.2% of the children's bedrooms had high levels of cockroach allergen in dust [47]. That study also found that children who were both allergic to cockroach allergen and exposed to high levels of this allergen had 0.37 hospitalizations a year, as compared with 0.11 for other children [47]. In areas heavily infested with cockroaches, constant exposure to house dust contaminated with cockroach allergens is unavoidable. Accordingly, many people become sensitized and develop cockroach allergy. As for management of cockroach allergy, recent research has shown that even just one pest control intervention (application of cockroach baits in homes) can lead to cockroach eradication and improve asthma outcomes in children [48].

Mayflies and caddisflies are delicate flies that spend most of their lives underwater as immatures. They emerge as adults in the spring and summer in tremendous numbers, are active for a few days, and then die (Fig. 1.19). They do not bite or sting, but body particles from mass emergence of these insects have been well documented as causing allergies.

**Fig. 1.19**   Mayfly adult. (Figure from USDA ARS Misc. Publ. No. 1443, 1986)

**Fig. 1.20**   Confused flour
beetle, larva, and adult.
(From USDA, ARS, Agri.
Hndbk. # 655, 1991)

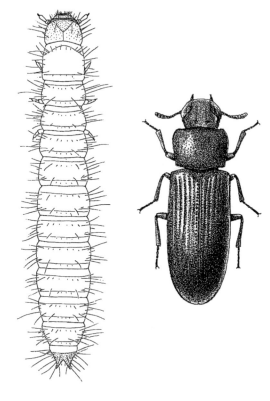

Nonbiting midges in the family Chironomidae have also been implicated as causes of insect inhalant allergy. A greater prevalence of asthma has been demonstrated in African populations seasonally exposed to the "green nimitti" midge, *Cladotanytarsus lewisi* [49, 50]. Kagen et al. [51] implicated *Chironomus plumosus* as a cause of respiratory allergy in Wisconsin.

In addition to inhalant allergens, adult beetles and larval flies, moths, or beetles, as well as their cast skins, often contaminate food and may be responsible for irritation and/or allergic responses through ingestion. The confused flour beetle, *Tribolium confusum*, and rice weevil, *Sitophilus granaries*, have been reported to cause allergic reactions in bakery workers (Fig. 1.20) [52]. In addition, physicians are often confronted with parents worried about their children who have inadvertently eaten a "maggot" in their cereal, candy bar, or other food products. These "maggots" may be moth, beetle, or fly larvae and generally cause no problems upon ingestion. However, some beetle larvae (primarily the family Dermestidae) found in stored food products possess minute barbed hairs (hastisetae) and slender elongate hairs (spicisetae) that apparently can cause enteric problems [53]. The symptoms experienced after ingesting dermestid larvae have been attributed to mechanical action of the hastisetae and spicisetae resulting in tissue damage or irritation in the alimentary tract. Clinical symptoms include diarrhea, abdominal pain, and perianal itch [54, 55].

Cockroaches seem to be most often involved in allergic responses from ingestion. Allergens are present in cockroach feces, which can be inadvertently ingested in heavily infested areas. Other allergens are present in cockroach saliva and exoskeletons which can be introduced into foodstuffs [52].

# References

1. Borror DJ, Triplehorn CA, Johnson NF. An introduction to the study of insects. 6th ed. Philadelphia: Saunders College Publishing; 1989.
2. Lane RP, Crosskey RW. Medical insects and arachnids. New York: Chapman and Hall; 1996. p. 723.
3. Herbert PDN, Ratnasingham S, deWaard JR. Barcoding animal life: cytochrome c oxidase subunit 1 divergences among closely related species. R Soc London Ser B. 2003;270:S96–S9.
4. Will KW, Rubinoff D. Myth of the molecule: DNA barcodes for species cannot replace morphology for identication and classification. Cladistics. 2004;20:47–55.
5. Sperling F. Butterfly molecular systematics: from species definitions to higher-lever phylogenies. In: Boggs CL, Watt WB, HEhrlich PR, editors. Butterflies: ecology and evolution taking flight. Chicago: University of Chicago Press; 2003. p. 431–58.
6. Bressler K, Shelton C. Ear foreign-body removal: a review of 98 consecutive cases. Laryngoscope. 1993;103:367–70.
7. Cushing E. History of entomology in World War II. Washington, DC: Smithsonian Institution; 1957. p. 117.
8. Mumcuoglu YK, Zias J. Head lice from hair combs excavated in Israel and dated from the first century BC to the eighth century AD. J Med Entomol. 1988;25:545–7.
9. Hoeppli R. Parasitic diseases in Africa and the Western Hemisphere: early documentation and transmission by the slave trade. Acta Tropica Suppl. 1969;10:33–46.
10. Harwood RF, James MT. Entomology in human and animal health. 7th ed. New York: Macmillan; 1979. p. 548.

11. Crosby MC. The American plague. New York: Berkley Books; 2006. p. 308.
12. Snetsinger R. The Ratcatcher's child: history of the pest control industry. Cleveland: Franzak and Foster Co.; 1983. p. 294.
13. Freedman B. Sanitarian's handbook. 4th ed. New Orleans: Peerless Publishing Co.; 1977.
14. Anonymous. An enduring reminder of the importance of public health. Lancet Infect Dis. 2018;18:1.
15. Barnard JH. Studies of 400 hymenoptera sting deaths in the United States. J Allergy Clin Immunol. 1973;52:259–64.
16. Frazier CA. Insect allergy: allergic reactions to bites of insects and other arthropods. Warren H. Green: St. Louis; 1969. p. 383.
17. Goddard J. Direct injury from arthropods. Lab Med. 1994;25:365–71.
18. Reisman RE. Insect stings. N Engl J Med. 1994;331:523–7.
19. O'Neil ME, Mack KA, Gilchrist J. Epidemiology of non-canine bite and sting injuries treated in U.S. emergency departments, 2001–2004. Public Health Rep. 2007;122:764–75.
20. Goddard J. Physician's guide to arthropods of medical importance. 6th ed. Boca Raton: Taylor and Francis (CRC); 2013. p. 412.
21. Simard JM, Watt DD. Venoms and toxins. In: Polis GA, editor. The biology of scorpions. Stanford: Stanford University Press; 1990. p. 414–44.
22. Alexander JO. Arthropods and human skin. Berlin: Springer-Verlag; 1984. p. 422.
23. Sherman RA. Maggot debridement in modern medicine. Inf Med. 1998;15:651–6.
24. Rossi AJ, Rossi A, Marin-Neto JA. Chagas' disease. Lancet. 2010;375:1388–402.
25. WHO. African trypanosomiasis: World Health Organization, Media Center., Fact Sheet Number 259. Geneva: Switzerland; 2010. p. 6.
26. Calisher CH. Persistent emergence of dengue. Emerg Infect Dis. 2005;11:738–9.
27. Gubler DJ. Epidemic dengue and dengue hemorrhagic fever: a global public health problem in the 21st century. In: Scheld WM, Armstrong D, Hughes JM, editors. Emerging infections, vol. 1. Washington, DC: ASM Press; 1998. p. 1–14.
28. Spira AM. Dengue: an underappreciated threat. Inf Med. 2005;22:304–6.
29. Enserink M. Malaria's miracle drug in danger. Science (News Focus). 2010;328:844–6.
30. Strausbaugh LJ. Emerging infectious diseases: a challenge to us all. Am Fam Phys. 1997;55:111–7.
31. Guerrant RL, Walker DH, Weller PF. Tropical infectious diseases. 3rd ed. Philadelphia: Saunders Elsevier; 2011. p. 1130.
32. CDC. Summary of notifiable infectious diseases and conditions -- United States, 2015. CDC, MMWR. 2017;64(53):1–144.
33. Paddock CD, Childs J. *Ehrlichia chaffeensis*: a prototypical emerging pathogen. Clin Microbiol Rev. 2003;16:37–64.
34. Hoffman DR. Allergic reactions to biting insects. In: Levine MI, Lockey RF, editors. Monograph on insect allergy. 2nd ed. Milwaukee: American Academy of Allergy and Immunology; 1986.
35. Moffitt JE, de Shazo RD. Allergic and other reactions to insects. In: Rich RR, Fleisher WT, Kotzin BL, Schroeser HW, editors. Rich's clinical immunology: principles and practice. 2nd ed. New York: Mosby; 2001.
36. Moffitt JE, Venarske D, Goddard J, Yates AB, deShazo RD. Allergic reactions to *Triatoma* bites. Ann Allergy Asthma Immunol. 2003;91(2):122–8; quiz 8–30, 94
37. Minocha R, Wang C, Dang K, Webb CE, Fernández-Peñas P, Doggett SL. Systemic and erythrodermic reactions following repeated exposure to bites from the common bed bug *Cimex lectularius* (Hemiptera: Cimicidae). Aust Entomol. 2016. https://doi.org/10.1111/aen.12250.
38. Gluck JC, Pacin MP. Asthma from mosquito bites: a case report. Ann Allergy. 1986;56:492–3.
39. Commins SP, James HR, Kelly LA, Pochan SL, Workman LJ, Perzanowski MS, et al. The relevance of tick bites to the production of IgE antibodies to the mammalian oligosaccharide galactose-alpha-1,3-galactose. J Allergy Clin Immunol. 2011;127(5):1286–93.
40. Platts-Mills TA, Schuyler AJ, Tripathi A, Commins SP. Anaphylaxis to the carbohydrate side chain alpha-gal. Immunol Allergy Clin N Am. 2015;35:247–60.
41. Ramey K, Stewart PH. Top ten facts you should know about "alpha-gal," the newly described delayed red meat allergy. J Miss State Med Assoc. 2016;57:279–81.

42. Gore JC, Schal C. Cockroach allergen biology and mitigation in the indoor environment. Annu Rev Entomol. 2007;52:439–63.
43. Spieksma FTM. The mite fauna of house dust, with particular reference to the house dust mite. Acarologia. 1967;9:226–34.
44. Spieksma FTM. The house dust mite, *Dermatophagoides pteronyssinus*, producer of house dust allergen. Thesis, University of Leiden, Netherlands, p. 65; 1967.
45. Htut T, Vickers L. The prevention of mite-allergic asthma. Int J Environ Health Res. 1995;5:47–61.
46. Cameron MM. Can house dust mite-triggered atopic dermatitis be alleviated using acaricides. Br J Dermatol. 1997;137:1–8.
47. Rosenstreich DL, Eggleston P, Kattan M, Baker D, Slavin RG, Gergen P, et al. The role of cockroach allergy and exposure to cockroach allergen in causing morbidity among inner-city children with asthma. N Engl J Med. 1997;336:1356–60.
48. Rabito FA, Carlson JC, He H, Werthmann D, Schal C. A single intervention for cockroach control reduces cockroach exposure and asthma morbidity in children. J Allergy Clin Immunol. 2017;140:565–70.
49. Gad el Rab MO, Kay AB. Widespread immunoglobulin E-mediated hypersensitivity in the Sudan to the "green nimitti" midge, *Cladotanytarsus lewisi*. J Allergy Clin Immunol. 1980;66:190–3.
50. Kay AB, MacLean CM, Wilkinson AH, Gad El Rab MO. The prevalence of asthma and rhinitis in a Sudanese community seasonally exposed to a potent airborne allergen, the "green nimitti" midge, *Cladotanytarsus lewisi*. J Allergy Clin Immunol. 1983;71:345–52.
51. Kagen SL, Yunginger JW, Johnson R. Lake fly allergy: incidence of chironomid sensitivity in an atopic population. J Allergy Clin Immunol. 1984;73:187.
52. Arlian LG. Arthropod allergens and human health. Annu Rev Entomol. 2002;47:395–433.
53. Lillie TH, Pratt GK. The hazards of ingesting beetle larvae. USAF Med Ser Dig. 1980;31:32.
54. Jupp WW. A carpet beetle larva from the digestive tract of a woman. J Parasitol. 1956;42:172.
55. Okumura GT. A report of canthariasis and allergy caused by *Trogoderma*. Calif Vect Views. 1967;14:19–20.

# Chapter 2
# Dynamics of Arthropod-Borne Diseases

## 2.1 Mechanical Versus Biological Transmission of Pathogens

Transmission of etiologic agents by arthropods is a complex phenomenon, and generalizations are difficult to make. Just because an arthropod feeds on a diseased host does not ensure that it can become infected nor does it ensure (even if disease agents are ingested) that ingested pathogens can survive and develop. There is considerable misunderstanding about this. When bitten by a tick, people think of Lyme disease (or something similar), often insisting that their physician prescribes an antibiotic prophylactically. Little do they realize that there are many tick species and not all are capable of disease transmission [1]. Further, they fail to realize that not every tick in nature (even within a vector species) is infected. Depending on the disease and area of the country, the presence of an infected tick can be like a needle in a haystack.

Arthropods capable of transmitting disease organisms to vertebrate hosts are called vectors [2]. For example, mosquitoes in the genus *Anopheles* are vectors of malaria organisms. Interestingly, no other mosquitoes are able to acquire and transmit the parasites. Other mosquitoes certainly feed on diseased humans but fail to become infected. Myriad factors affect the ability of arthropods to acquire, maintain, and ultimately, transmit pathogens. An understanding of arthropod—pathogen interactions—is crucial to preventing and/or managing vector-borne diseases. First, a distinction must be made between mechanical and biological transmission and their various modes (Table 2.1).

### 2.1.1 *Mechanical Transmission*

Mechanical transmission of disease agents occurs when arthropods physically carry pathogens from one place or host to another host—often via body parts. For example, flies and cockroaches have numerous hairs, spines, and setae on their bodies

© Springer International Publishing AG, part of Springer Nature 2018
J. Goddard, *Infectious Diseases and Arthropods*, Infectious Disease,
https://doi.org/10.1007/978-3-319-75874-9_2

**Table 2.1** Modes of pathogen/parasite transmission[a]

| Mode of transmission | Example |
| --- | --- |
| Mechanical transmission | Pathogens on cockroach body parts |
| Biological transmission | |
|    Transmission by eating vector | Fleas—Dog tapeworm |
|    Transmission during/after bloodsucking | |
|    Proliferation in gut and transmission in feces | Kissing bugs—Chagas' disease |
|    Proliferation in gut and transmission by bite | Fleas—Plague |
|    Penetration of gut and transmission by bite | Mosquitoes—Malaria |

[a]Adapted from Lane and Crosskey [3]

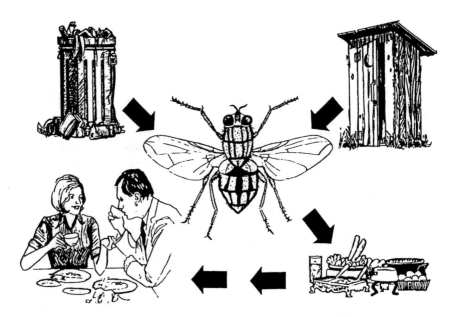

**Fig. 2.1** Example of mechanical transmission of disease agents. (CDC figure)

that collect contaminants as the insects feed on dead animals or excrement (Figs. 2.1 and 2.2). When they subsequently walk on food or food preparation surfaces, mechanical transmission occurs [4–6]. Mechanical transmission may also occur if a blood-feeding arthropod has its feeding event disrupted. For example, if a mosquito feeds briefly on a viremic bird and is interrupted, a subsequent immediate feeding on a second bird could result in virus transmission. This would be similar to an accidental needle stick. The main point about mechanical transmission is that the pathogen undergoes no development (cyclical changes in form and so forth) and no significant multiplication. It is just there for the ride.

**Fig. 2.2** House flies are major mechanical transmitters of disease agents. (Photo copyright 2011 by Jerome Goddard Ph.D.)

## 2.1.2 Biological Transmission

In biological transmission, there is either multiplication or development of the pathogen in the arthropod or both [7, 8]. Table 2.2 provides a detailed list of many of these vector-borne pathogens. Biological transmission may be classified into three types. In *cyclodevelopmental transmission*, the pathogen must undergo a cycle of development within the arthropod vector, but no multiplication. For example, the filarial worm causing bancroftian filariasis, when first ingested by mosquitoes, is not infective to a vertebrate host—it must undergo a period of development. *Propagative transmission* means the pathogen must multiply before transmission can occur. There is no cyclical change or development of the organism—plague bacteria in fleas, for example. Finally, in *cyclopropagative transmission*, the pathogen must undergo both cyclical changes and multiplication. The classical example of this is malaria plasmodia in *Anopheles* mosquitoes.

Biological transmission reflects an evolutionary adaptation of the parasite into a cyclic event between vertebrate host and arthropod vector. This involves several factors, including the arthropod feeding on the right host, feeding in such a way (or time) that the parasites, circulating in the peripheral blood of the host animal, are ingested, and a mechanism for getting into a new host—often by penetrating the gut wall of the arthropod and subsequently migrating to a site for reinjection. All of this becomes a fine-tuned system operating efficiently for countless generations.

Take plague as an example of the complex interplay of factors affecting disease transmission [2]. *Yersinia pestis*, the causative agent, is essentially a disease of rodents that occasionally spills over into the human population (Fig. 2.3). The enzootic cycle (established, ongoing among animals) is primarily mechanical, with the rodent hosts being relatively resistant. In the epizootic cycle (occasional outbreaks or epidemics), susceptible rodent populations become infected, resulting in mass die-offs. Fleas on epizootic hosts become heavily infected with bacilli and regurgitate

**Table 2.2** Arthropod-borne or caused human illnesses

| Disease | Pathogen | Type | Primary vector | Common name |
|---|---|---|---|---|
| Yellow fever | *Flavivirus* | Virus | *Aedes aegypti, A. africanus* | Yellow fever mosquito |
| Dengue fever | *Flavivirus* | Virus | *Aedes aegypti* | Yellow fever mosquito |
| Malaria | *Plasmodium* spp. | Protozoan | *Anopheles* spp. | Mosquito |
| Filariasis | *Wuchereria bancrofti*, others | Nematode | *Anopheles* and *Culex* spp. | Mosquito |
| Rift Valley fever | *Phlebovirus* | Virus | *Aedes* and *Culex* spp. | Mosquito |
| West Nile virus | *Flavivirus* | Virus | *Culex pipiens, C. quinquefasciatus* | Northern/southern house mosquito |
| St. Louis encephalitis | *Flavivirus* | Virus | *Culex pipiens, C. quinquefasciatus* | Northern/southern house mosquito |
| Eastern equine encephalitis | *Flavivirus* | Virus | *Culiseta melanura* (enzootic), *Coquillettidia* sp., *Aedes* spp. (epizootic) | Mosquito |
| La Crosse encephalitis | *Bunyavirus* | Virus | *Aedes triseriatus* | Tree-hole mosquito |
| African sleeping sickness (human African trypanosomiasis) | *Trypanosoma brucei gambiense, T. brucei rhodiense* | Protozoan | *Glossina* spp. | Tsetse fly |
| Epidemic relapsing fever | *Borrelia recurrentis* | Bacterium | *Pediculus humanus* | Body louse |
| Epidemic typhus | *Rickettsia prowazekii* | Rickettsia | *Pediculus humanus* | Body louse |
| Trench fever | *Bartonella quintana* | Bacterium | *Pediculus humanus* | Body louse |
| Leishmaniasis | *Leishmania donovani, L. braziliensis*, others | Protozoan | *Phlebotomus, Lutzomyia* spp. | Sand fly |
| Sand fly fever | *Phlebovirus* | Virus | *Phlebotomus* | Sand fly |
| Onchocerciasis "river blindness" | *Onchocerca volvulus* | Nematode | *Simulium* spp. | Black fly |
| Endemic (murine) typhus | *Rickettsia typhi* | Rickettsia | *Xenopsylla cheopis* | Rat flea |
| Plague | *Yersinia pestis* | Bacterium | *Xenopsylla cheopis* | Rat flea |
| Tularemia | *Francisella tularensis* | Bacterium | *Chrysops discalis, Dermacentor variabilis, D. andersoni* | Deer fly, tick |
| Cutaneous anthrax | *Anthracis bacillus* | Bacterium | *Chrysops* spp. | Deer fly |
| Loiasis | *Loa loa* | Nematode | *Chrysops silacea, C. dimidiata* | Deer fly, mango fly |
| Chagas' disease | *Trypanosoma cruzi* | Protozoan | Triatominae | Kissing bug |

(continued)

**Table 2.2** (continued)

| Disease | Pathogen | Type | Primary vector | Common name |
|---------|----------|------|----------------|-------------|
| Tick-borne relapsing fever | *Borrelia* spp. | Spirochete | *Ornithodoros turicata, O. hermsii, O. parkeri,* others | Soft tick |
| Babesiosis | *Babesia microti* and others | Protozoa | *Ixodes scapularis* | Deer tick |
| Colorado tick fever | *Reovirus* | Virus | *Dermacentor andersoni* | Rocky Mountain wood tick |
| Ehrlichiosis | *Ehrlichia chaffeensis, E. ewingii, E. muris*-like agent | Bacterium | *Amblyomma americanum, Ixodes scapularis* | Lone star tick, deer tick |
| Anaplasmosis | *Anaplasma phagocytophilum* | Bacterium | *Ixodes scapularis* | Deer tick |
| Lyme disease | *Borrelia burgdorferi* | Bacterium | *Ixodes scapularis, I. pacificus* | Deer tick |
| Q fever | *Coxiella burnetii* | Rickettsia | Many tick species | Hard tick |
| Rocky Mountain spotted fever | *Rickettsia rickettsii* and related species | Rickettsia | *Dermacentor andersoni, D. variabilis, Rhipicephalus sanguineus* | Rocky Mountain wood tick, American dog tick, brown dog tick |
| Tick-borne encephalitis | Togavirus | Virus | Primarily *Ixodes* spp. | Hard tick |
| Rickettsialpox | *Rickettsia akari* | Rickettsia | *Liponyssoides sanguineus* | House mouse mite |
| Scabies | – | – | *Sarcoptes scabiei* | Mite |
| Scrub typhus | *Orientia tsutsugamushi* | Rickettsia | *Leptotrombidium* spp. | Chigger mite |

into feeding wounds. There may also be other modes of transmission during epizootics, such as cats eating infected rodents, becoming pneumonic, and directly infecting humans by coughing. Obviously, the worst-case transmission scenario is development of primary pneumonic plague in humans (transmission by coughing), resulting in a widespread outbreak with tremendous case numbers.

Since vector-borne diseases are dynamic and quite complicated, basic research into arthropod vectorial capacity is of great importance. Here basic research tremendously aids the medical community. By identifying animal hosts and which arthropod species are "competent" vectors (*see* Sect. 2.2) and targeting control measures toward those species, disease transmission can be interrupted, leading to abatement of the epidemic. Interruption of the transmission cycle is especially important for viral diseases (e.g., mosquito-borne encephalitis), which have no specific treatments. I personally have been involved in eastern equine encephalitis outbreaks where the only hope of stopping the appearance of new cases was to identify the vector species in the area and direct specific mosquito control measures toward them.

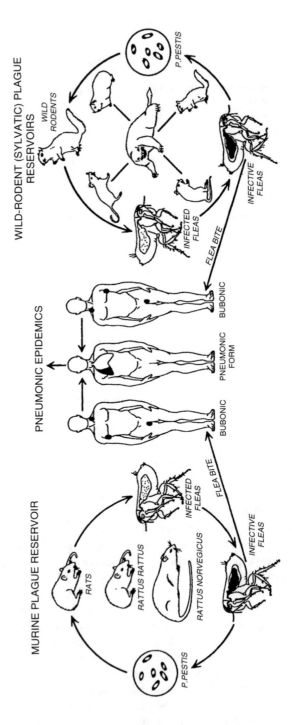

**Fig. 2.3** Plague life cycle. (CDC figure)

## 2.2   Vector Competence

Vector competence refers to the ability of arthropods to acquire, maintain, and transmit microbial agents [1]. As mentioned, not all arthropods are vectors of disease agents. Even blood-feeding arthropods may not be vectors. Bed bugs are a great example of this—they feed on humans who may have a variety of disease agents but apparently do not become infected [9]. Insects, ticks, or mites may "pick up" a pathogen with their blood meal, but the pathogen must overcome many obstacles before being transmitted to another host. In many cases, the gut wall must be bypassed; the pathogen must survive (and even develop) in arthropod tissues, such as hemolymph, muscles, or the reproductive system, and finally, must penetrate the salivary glands for injection into a new host. Note: in some cases, transmission occurs without the pathogen making its way into the salivary glands (*see* Table 2.1). In the meantime, the arthropod itself must live long enough for all of this multiplication/movement/development to take place. An ideal vector then would be one providing a suitable internal environment for the pathogen, be long-lived, have a host feeding pattern matching the host range of the pathogen, feed often and for extended periods, ingest large amounts of blood in each life stage, and disperse readily [2]. Of course, no arthropod possesses all these characteristics, but some have varying degrees of them. In a specific region or season, there are primary vectors, which are the main arthropods involved in the transmission cycle of a given disease, and secondary vectors, which play a supplementary role in transmission, but would be unable to maintain the disease in the absence of primary vectors [8].

Both intrinsic and extrinsic factors affect vector competence. Intrinsic factors include internal physiological factors and innate behavioral traits governing infection of a vector and its ability to transmit an agent—things like duration of feeding, host preferences, whether or not there is transovarial transmission, and so forth. Extrinsic factors include number of host animals, their activity patterns, climatic conditions, genetic variation in the infectivity of the pathogen, and so on. Competition between microorganisms inside a vector may also affect vector competence. This has often been referred to as the "interference phenomenon" [1, 8, 10]. A good example occurs in ticks. Burgdorfer et al. [11] reported that the tick *Dermacentor andersoni* from the east side of the Bitterroot Valley in western Montana contained a nonpathogenic spotted fever group (SFG) rickettsia, which they named the east side agent. East side agent was eventually described as a new species, *Rickettsia peacockii* [12]. This rickettsia, closely related to the causative agent of Rocky Mountain spotted fever (RMSF), *Rickettsia rickettsii*, is rarely present in tick blood (hemolymph) and is readily missed by the standard tick testing method—the hemolymph test. The rickettsiae are confined primarily to portions of the tick's midgut and, most importantly, the ovaries. *R. peacockii* is maintained in the tick population through transovarial transmission, and infected ticks are refractory to ovarian infection with *R. rickettsii*. However, these ticks are susceptible to experimental infection with *R. rickettsii* and may transmit the infection horizontally (stage to stage). Thus, ticks infected with *R. peacockii* and infected experimentally with *R. rickettsii* are unable to transmit *R. rickettsii* to their progeny. In effect, infection of the tick, *D. andersoni*, with *R. peacockii*, blocks the subsequent ability of the ticks to transmit *R. rickettsii* transovarially. Other experiments have also demonstrated that tick ovarial infection with one rickettsial species precludes secondary infection with other rickettsiae [13]. This "interference phenomenon" provides an explanation for the curious long-standing disease situation in the Bitterroot Valley. Most cases of RMSF have occurred on the west side of the valley where *D. andersoni* is abundant; on the east side, *D. andersoni* is also abundant and is reported to bite local residents, yet few locally acquired cases occur there. With *R. peacockii* in the area, *R. rickettsii* cannot be maintained transovarially—it can only be maintained transstadially. Thus, long-term maintenance cannot be sustained. Burgdorfer et al. [11] say that transovarial interference of *R. rickettsii* in *D. andersoni* ticks may also be mediated by other nonpathogenic SFG rickettsia, such as *Rickettsia montana* and *Rickettsia rhipicephali*. Most ticks in nature infected with rickettsial organisms harbor nonpathogenic species. Thus, transovarial interference may have epidemiologic significance—it may explain why ticks collected from various geographic regions are not infected with two or more species of SFG rickettsiae [10].

### 2.2.1  Incrimination of Vectors: A Complicated Issue

To illustrate the difficulty in incriminating vectors of a specific disease, the following discussion on malaria in the western United States is provided as an example. Much of this discussion is from McHugh [14], Porter and Collins [15], and Jensen et al. [16].

Concerning malaria in the western United States, we must first consider what criteria the mosquito must fulfill to be proven to be the primary, or at least an important, vector of the human malaria parasites:

- It must be a competent vector of the parasites.
- Its geographic distribution must match the transmission pattern.
- It must be abundant.
- It must be anthropophilic.
- It must be long lived.
- Field collections should demonstrate a measurable proportion of the mosquito.
- Population infected (usually about 1%).

*Anopheles freeborni* (*sensu latu*) is certainly well known through laboratory transmission studies as a competent vector of a number of *Plasmodium* species including *Plasmodium falciparum* from Panama and Zaire, *Plasmodium vivax* from Vietnam, and *Plasmodium malariae* from Uganda to name a few. Does this mean *An. freeborni* is a vector of those malarial parasites in those areas? Of course not, the mosquito does not occur there. That is, it does not fulfill the second criterion.

What about *An. freeborni* in the western United States? Because this species is a competent vector, is widely distributed, and is often abundant, it is frequently cited as the most likely suspect vector. However, it turns out that *An. freeborni* is a generalist feeder and not particularly anthropophilic. Several studies in California found only 1–3% of several thousand field-caught females had fed on humans. Longevity studies of this species indicated a daily survivorship of about 0.72–0.74 for female *An. freeborni*. Based on this estimate, an initial infected blood meal on day 3 postemergence, and an extrinsic incubation period of about 12 d, the probability of a female living long enough to be infective would be on the order of 0.0072 or less.

What about the last criterion—finding infected mosquitoes in field collections? There have been only a very limited number of isolations of any human malaria parasite from any species of *Anopheles* collected in the western United States. Dr. Bill Reeves at UC Berkeley gave an anecdotal report of oocysts on the gut of *An. freeborni* collected in California during the mid-1940s, and he also reported infected *An. freeborni* from New Mexico at that same time. However, as will be discussed below, changes in nomenclature and our understanding of mosquito systematics, not to mention the failure to provide a specific determination of the parasites involved, make it impossible to ascribe much significance to these reports.

Considering these data, particularly host selection (i.e., low rate of human feedings) and survivorship (i.e., low), *An. freeborni* may be overrated as a potential vector. Perhaps another species may be responsible—such as *Anopheles punctipennis*. If one visits a number of locations where autochthonous cases of malaria have occurred in California, he or she will be struck by the fact that most cases were acquired in riparian settings. This habitat is more typical of *An. punctipennis*. It turns out that Gray back in the 1950s published several insightful reviews drawing the same conclusion [17, 18]. Gray reported that *An. punctipennis* was actually more common than *An. freeborni* at the site of the famous Lake Vera outbreak of malaria in the early 1950s and was the probable vector. More recent evidence supports his claim [16].

There remain two problems in understanding the confusing epidemiology of malaria in California and, perhaps, the rest of the United States. Anthropogenic changes in the local ecology—damming and channeling rivers, introduction of the rice culture, destruction of riparian habitat, and so forth—have dramatically altered the landscape over the past 100 years. Thus, the mosquito species responsible for transmission may have changed over time. Second, the eastern US malaria vector, *Anopheles quadrimaculatus*, is actually a complex of several sibling species. Researchers at the USDA-ARS lab in Gainesville, FL, helped determine that *An. quadrimaculatus* (the vector in the eastern United States) is a complex of at least five identical-looking species. This may be the case with *An. freeborni* in the West. The late Ralph Barr and coworkers determined that what appeared to be *An. freeborni* collected in several sites of malarial transmission in Southern California were, in fact, a new species that they named in honor of W. B. herms (*Anopheles hermsi*). Therefore, it may be that earlier workers who suspected *An. freeborni* were correct to the extent that their technology (i.e., morphologic identifications) was capable of identifying the insects involved. Without access to mosquitoes collected in the past, especially those from early studies in which mosquitoes were still lumped as *Anopheles maculipennis*, it will be very difficult to determine what species were actually being studied. (As an aside, it would be very interesting to study extant laboratory colonies of "*An. freeborni*" and determine exactly which species are really being maintained and studied.)

We can draw two conclusions. First, the epidemiology/ecology of malaria is dynamic and may have changed over time, but the most likely vectors in the western United States at the present time are *An. hermsi*, *An. freeborni*, or *An. punctipennis*, with other species involved if conditions are appropriate. Second, to incriminate a specific vector, we must carefully consider the ecology of malarious foci and weigh all the factors that make an arthropod a good vector, not just focusing in on one or two.

# References

1. Lane RS. Competence of ticks as vectors of microbial agents with an emphasis on *Borrelia burgdorferi*. In: Sonenshine DE, Mather TN, editors. Ecological dynamics of tick-borne zoonoses. New York: Oxford University Press; 1994. p. 45–67.
2. McHugh CP. Arthropods: vectors of disease agents. Lab Med. 1994;25:429–37.
3. Lane RP, Crosskey RW. Medical insects and arachnids. New York: Chapman and Hall; 1996. p. 723.
4. Bressler K, Shelton C. Ear foreign-body removal: a review of 98 consecutive cases. Laryngoscope. 1993;103:367–70.
5. Kopanic RJ, Sheldon BW, Wright CG. Cockroaches as vectors of *Salmonella:* laboratory and field trials. J Food Prot. 1994;57:125–32.
6. Zurek L, Schal C. Evaluation of the German cockroach as a vector for verotoxigenic *Escherichia coli* F18 in confined swine production. Vet Microbiol. 2004;101:263–7.
7. Chamberlain RW, Sudia WD. Mechanism of transmission of viruses by mosquitoes. Annu Rev Entomol. 1961;6:371–90.
8. Harwood RF, James MT. Entomology in human and animal health. 7th ed. New York: Macmillan; 1979. p. 548.
9. Goddard J, de Shazo RD. Bed bugs (*Cimex lectularius*) and clinical consequences of their bites. J Am Med Assoc. 2009;301:1358–66.
10. Azad AF, Beard CB. Rickettsial pathogens and their arthropod vectors. Emerg Infect Dis. 1998;4:179–86.
11. Burgdorfer W, Brinton LP. Mechanisms of transovarial infection of spotted fever rickettsiae in ticks. Ann N Y Acad Sci. 1975;266:61–72.
12. Niebylski ML, Peacock MG, Schwan TG. Lethal effect of *Rickettsia rickettsii* on its tick vector (*Dermacentor andersoni*). Appl Environ Microbiol. 1999;65:773–8.
13. Macaluso KR, Sonenshine DE, Ceraul SM, Azad AF. Rickettsial infection in *Dermacentor variabilis* inhibits transovarial transmission of a second rickettsia. J Med Entomol. 2002;39:809–13.
14. McHugh CP. Ecology of a semi-isolated population of adult *Anopheles freeborni*: abundance, trophic status parity, survivorship, gonotrophic cycle length, and host selection. Am J Trop Med Hyg. 1989;41:169–76.
15. Porter CH, Collins FH. Susceptibility of *Anopheles hermsi* to *Plasmodium vivax*. Am J Trop Med Hyg. 1990;42:414–6.
16. Jensen T, Dritz DA, Fritz GN, Washino RK, Reeves WC. Lake Vera revisited: parity and survival rates of *Anopheles punctipennis* at the site of a malaria outbreak in the Sierra Nevada foothills of California. Am J Trop Med Hyg. 1998;59:591–4.
17. Gray HF. The confusing epidemiology of malaria in California. Am J Trop Med Hyg. 1956;5:411–8.
18. Gray HF, Fontaine RE. A history of malaria in California. Proc Calif Mosq Control Assoc. 1957;25:1–20.

# Part II
# Major Arthropod-Borne Diseases

# Chapter 3
# Mosquito-Borne Diseases

## 3.1 Basic Mosquito Biology

Mosquitoes are flies and, thus, undergo complete metamorphosis, having egg, larval, pupal, and adult stages (Fig. 3.1). Larvae are commonly referred to as wigglers and pupae as tumblers. Larvae and pupae of mosquitoes are always found in water. Breeding sites may be anything from water in discarded automobile tires to water in the axils of plants and to children's toys, pools, puddles, swamps, and lakes. Mosquito species differ greatly in their breeding habits, biting behavior, flight range, and so forth (Fig. 3.2). However, a generalized description of their life cycle is presented here as a basis for understanding mosquito biology and ecology. There are two subfamilies in the mosquito family Culicidae—Anophelinae and Culicinae. Members of one tribe, Toxorhynchitini, in the Culicinae, are huge, non-bloodsucking mosquitoes whose larvae eat mosquito larvae of other species. The larvae have a breathing tube (siphon), but it is short and conical. Most larvae in the subfamily Culicinae hang down just under the water surface by the siphon, whereas anopheline larvae lie horizontally just beneath the water surface supported by small notched organs of the thorax and clusters of float hairs along the abdomen (Fig. 3.3). They have no prominent siphon. Mosquito larvae feed on suspended particles in the water as well as microorganisms. They undergo four molts (each instar successively larger), the last of which results in the pupal stage. With optimal food and temperature, the time required for larval development can be as short as 4–5 d.

Unlike most insect pupae, mosquito pupae are quite active and quickly swim (tumble) toward the bottom of their water source on disturbance. Pupae do not feed. They give rise to adult mosquitoes in 2–4 d. The emergence process begins with splitting of the pupal skin along the back. Upon eclosion, an adult must dry its wings and groom its head appendages before flying away (Fig. 3.4). Accordingly, this is a critical stage in the survival of mosquitoes. If there is too much wind or wave action, the emerging adult will fall over, becoming trapped on the water

© Springer International Publishing AG, part of Springer Nature 2018
J. Goddard, *Infectious Diseases and Arthropods*, Infectious Disease,
https://doi.org/10.1007/978-3-319-75874-9_3

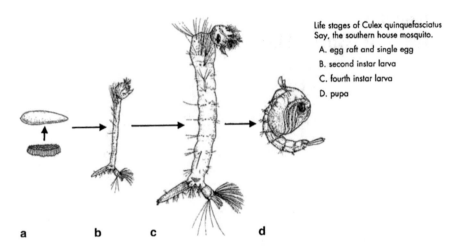

Life stages of *Culex quinquefasciatus* Say, the southern house mosquito.

A. egg raft and single egg

B. second instar larva

C. fourth instar larva

D. pupa

a            b            c                    d

**Fig. 3.1** Life stages of a mosquito. (From the Mississippi State University Extension Service, by Joe MacGown, with permission)

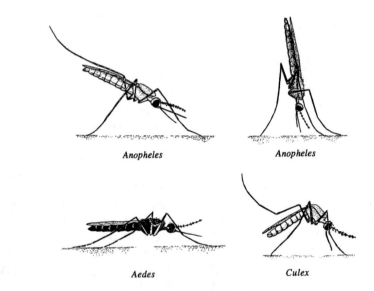

*Anopheles*                        *Anopheles*

*Aedes*                            *Culex*

**Fig. 3.2** Adult mosquitoes assume various positions, depending on the particular genus. (From E. Boles, The Mosquito Book, Mississippi Department of Health)

surface to die soon. This is the reason that little if any mosquito breeding occurs in open water; it occurs at the water's edge among weeds.

Adult mosquitoes of both sexes obtain nourishment for basic metabolism and flight by feeding on nectar. In addition, females of most species need a blood meal from birds, mammals, or other vertebrates for egg development. They suck blood via specialized piercing–sucking mouthparts (Fig. 3.5). Breeding sites selected for egg laying differ by species, but generally mosquitoes can be divided into three major groups: permanent water breeders, floodwater breeders, and artificial container/tree

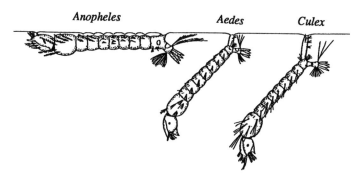

**Fig. 3.3** *Culex* and *Aedes* mosquitoes breathe via a siphon tube, whereas *Anopheles* mosquitoes do not. (From E. Boles, The Mosquito Book, Mississippi Department of Health)

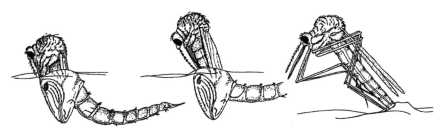

**Fig. 3.4** Adult mosquitoes emerging from pupal stage—a critical stage in development. (From E. Boles, The Mosquito Book, Mississippi Department of Health)

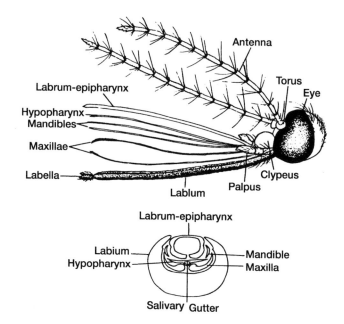

**Fig. 3.5** Head and mouthparts, with a cross section of proboscis, of female mosquito. (From USDA technical bulletin, No. 1447)

hole breeders. *Anopheles* and many *Culex* mosquitoes select permanent water bodies, such as swamps, ponds, lakes, and ditches that do not usually dry up. Floodwater mosquitoes lay eggs on the ground in low areas subject to repeated flooding. During heavy rains, water collecting in these low areas covers the eggs, which hatch from within minutes to a few hours. Salt-marsh mosquitoes (*Aedes sollicitans*), inland floodwater mosquitoes (*Aedes vexans*), and dark rice field mosquitoes (*Psorophora columbiae*) are included in this group. Artificial container/tree hole breeders are represented by yellow fever mosquitoes (*Aedes aegypti*), Asian tiger mosquitoes (*Aedes albopictus*), tree hole mosquitoes (*Aedes triseriatus*), and others. However, several species of *Anopheles* and *Culex* may also occasionally oviposit in these areas. Some of these container-breeding species lay eggs on the walls of a container just above the waterline. The eggs are flooded when rainfall raises the water level. Other species oviposit directly on the water surface.

Female *Anopheles* mosquitoes generally lay eggs on the surface of the water at night. Each batch usually contains 100–150 eggs. Each *Anopheles* egg is cigar-shaped, about 1 mm long, and bears a pair of air-filled floats on the sides. Under favorable conditions, hatching occurs within 1 or 2 d. *Anopheles* mosquitoes may occur in extremely high numbers. In the Mississippi Delta, mosquito trapping has yielded as many as 9000 *Anopheles quadrimaculatus*/trap/night!

*Aedes* mosquitoes lay their eggs on moist ground around the edge of the water or, as previously mentioned, on the inside walls of artificial containers just above the waterline. *Aedes* eggs will desiccate and perish easily when first laid. However, after embryo development with each egg, the eggs can withstand dry conditions for long periods of time. This trait has allowed *Aedes* mosquitoes to use temporary water bodies for breeding, such as artificial containers, periodically flooded salt marshes or fields, tree holes, and storm water pools. Also, *Aedes* mosquitoes have inadvertently been carried to many parts of the world as dry eggs in tires, water cans, or other containers. The Asian tiger mosquito (*Ae. albopictus*) was introduced into the United States in the 1980s in shipments of used truck tire casings imported from Taiwan and Japan. Once these tires were stacked outside and began to collect rainwater, the eggs hatched.

Salt-marsh mosquitoes, such as *Aedes taeniorhynchus* and *Ae. sollicitans*, breed in salt-marsh pools flooded by tides and/or rain and periodically emerge in great swarms, making outdoor activity in large areas of seacoast unbearable. Their flight range is between 5 and 10 miles, but they can travel with the wind 40 miles or more. *Psorophora* mosquitoes also lay dry-resistant eggs. These mosquitoes are a major problem species in rice fields. Eggs are laid on the soil and hatch once the rice field is irrigated. *Psorophora* mosquitoes may also emerge in huge swarms. In 1932, *Psorophora columbiae* is reported to have caused a great loss of livestock in the Everglades, and the milk supply was greatly reduced during the 4 d of the infestation [1].

*Culex* mosquitoes lay batches of eggs attached together to form little floating rafts. On close inspection of a suitable breeding site, these egg rafts can often be seen floating on the water's surface. Where breeding conditions are favorable, *Culex* mosquitoes also occur in enormous numbers. Several *Culex* species are notorious for their aggravating high-pitched hum when flying about the ears.

In tropical areas, mosquito breeding may continue year round, but in temperate climates, many species undergo a diapause in which the adults enter a dormant state similar to hibernation. In preparation for this, females become reluctant to feed, cease ovarian development, and develop fat body. In addition, they may seek a protected place to pass the approaching winter. Some species, instead of passing the winter as hibernating adults, produce dormant eggs or larvae that can survive the harsh effects of winter.

Mosquitoes vary in their biting patterns. Most species are nocturnal in activity, biting mainly in the early evening. However, some species, especially *Ae. aegypti* and *Ae. albopictus*, bite in broad daylight (although there may be a peak of biting very early and late in the day). Others, such as salt-marsh species and many members of the genus *Psorophora*, do not ordinarily bite during the day but will attack if disturbed (such as walking through high grass harboring resting adults).

## 3.2  Malaria

### 3.2.1  Introduction

Malaria is one of the most serious human diseases in the world. More people have probably died from malaria than from any other infectious disease in human history [2]. Although likely underreported, estimates of the annual number of clinical cases are approximately 250 million, with about 500,000 deaths—mainly in children [3–5]. Malaria generally occurs in areas of the world between 45°N and 40°S latitude. Although many countries are not entirely malarious, the WHO estimates that about 3.2 billion persons—well over 40% of the world's population—live in malarious areas [6]. The geographic distribution of malaria has shrunk significantly over the last 150 yr, mainly from eradication efforts in temperate zones (Figs. 3.6 and 3.7). However, it is fairly easy to eradicate the disease at the fringes of its geographic distribution and/or island locations. Although indigenous malaria disappeared from the United States in the 1950s, there have been several episodes of introduced malaria and subsequent autochthonous cases in this country over the last few decades. Introduced malaria occurs when local people are infected as a result of imported cases (travelers and so forth) or people having relapses from former cases. Overall, the malaria situation worldwide is improved, but there are several complicating factors such as mosquito vectors becoming resistant to many of the pesticides being used to control them, and in many areas the malaria parasites are resistant to the prophylactic drugs used to prevent the disease. In addition, civil strife and large-scale refugee movements are widespread in sub-Saharan Africa, and there is increased travel by nonimmune expatriates.

**Fig. 3.6** Approximate geographic distribution of malaria—1850

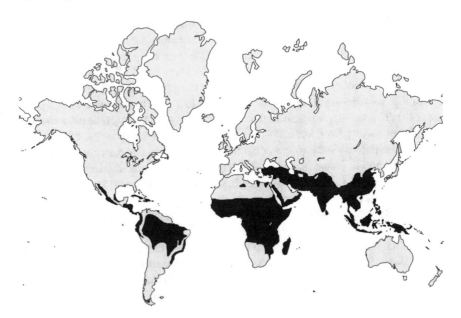

**Fig. 3.7** Approximate geographic distribution of malaria—current

### 3.2.2   The Disease and its Diagnosis

Classic malaria includes such symptoms as fever, chills, sweats, headache, muscle pain, and malaise. There may be a repeating cycle of high fever and sweating. Infants may display only lethargy, irritability, and anorexia. In rare forms of falciparum malaria (e.g., cerebral malaria), chills and fever may be absent, and the patient may present with medical shock, delirium, or coma [7, 8]. Falciparum malaria may also produce complications, such as renal failure, hemolytic anemia, hypoglycemia, and acute pulmonary edema. For this reason, falciparum malaria is often called pernicious or malignant malaria.

Diagnosis of malaria is frequently based on clinical presentation. Definitive "gold standard" diagnosis of malaria depends on identifying parasites in the blood. Both thick and thin blood smears need to be carefully examined by lab or parasitology personnel. It must be noted that a patient can be very sick and yet demonstrate very few parasites in blood smears. Repeated thick blood smears may be necessary every 2–6 h before the parasites are found. Rapid diagnostic tests (RDTs) for malaria are gaining increased use worldwide and are very useful in field studies and in remote tropical locations. These immunochromatographic tests are based on the capture of parasite antigens from peripheral blood. The US Food and Drug Administration has approved at least one RDT for use in the United States.

### 3.2.3   The Causative Agent and Life Cycle

Human malaria is caused by any one of five species of microscopic protozoan parasites in the genus *Plasmodium*—*Plasmodium vivax, Plasmodium malariae, Plasmodium ovale, Plasmodium falciparum*, and *Plasmodium knowlesi*. The infective sporozoites are transmitted to humans only by mosquitoes in the genus *Anopheles*. However, not every species of *Anopheles* is a vector; less than half of the more than 400 known species are considered vectors. In fact, only 45–50 species are important vectors. Not all species of *Plasmodium* occur in all places. Generally, *P. vivax* is prevalent throughout all malarious areas, except sub-Saharan Africa; *P. ovale* is found chiefly in tropical areas of Western Africa (occasionally Western Pacific and Southeast Asia); *P. malariae* is widely distributed around the world but often spotty; *P. falciparum* predominates in sub-Saharan Africa but is also common in Southeast Asia and South America; and *P. knowlesi* occurs in Southeast Asia. Overall, the predominant malaria species in the world is *P. falciparum*.

The malaria parasite life cycle is quite complicated and fraught with technical terms (Fig. 3.8). Only a brief summary will be presented here. Sporozoites injected during mosquito biting infect liver cells. After a time of growth, development, and division, merozoites are released from the liver into the bloodstream. There the parasites invade human red blood cells, where they grow and multiply asexually. After 48–72 h, the red blood cells burst, releasing large numbers of new parasites, most of which enter new red blood cells (this reinitiates the cycle). Other than these asexual forms, some of the parasites develop into

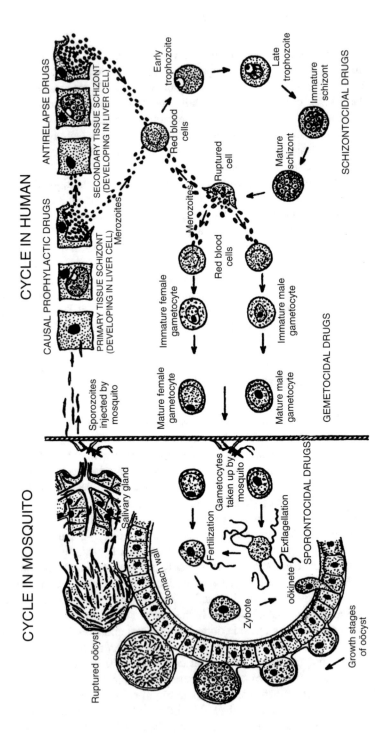

**Fig. 3.8** Malaria life cycle. (From US Navy, ref. [2])

sexual forms—male and female gametocytes. If a susceptible feeding *Anopheles* mosquito draws up gametocytes with its blood meal, fertilization takes place in the stomach. The resulting zygote penetrates the mosquito gut wall and forms an oocyst on the basement membrane of the gut. Eventually, oocysts rupture, releasing sporozoites inside the mosquito body cavity. After migration to the salivary glands, the mosquito is infective. The entire developmental time within the mosquito is 8–35 d.

### *3.2.4 Mosquito Vectors and Behavior*

As discussed in Chap. 2, biological transmission of any disease agent reflects an evolutionary adaptation of the parasite into a cyclic event between vertebrate host and arthropod vector. This involves several things, including the mosquito feeding on the right host, feeding in such a way (or time) that the parasites, circulating in the blood of the host animal, are ingested, and a mechanism for penetrating the gut wall of the mosquito and subsequently migrating to the salivary glands for reinjection into another host. All of this becomes a fine-tuned system that operates efficiently for countless generations. A highly efficient mosquito vector of malaria is one that is highly susceptible to the full development of the parasite (*Plasmodium*), prefers to feed on humans, and lives for a relatively long time [9].

Some notable malaria vectors worldwide are as follows: Several members of the *Anopheles gambiae* complex (consisting of seven almost identical species) are the most efficient malaria transmitters in Africa (Figs. 3.9 and 3.10). They often breed in freshwater exposed to sunlight. *Anopheles darlingi* is one of the major contributors to endemic malaria in extreme southern Mexico and Central and South America (Figs. 3.11 and 3.12). It breeds in shaded areas of freshwater marshes, swamps, lagoons, lakes, and ponds. The *Anopheles leucosphyrus* group (containing at least 20 closely related species) contains several main vectors of malaria in Southeast Asia (Figs. 3.13 and 3.14). They mostly breed in freshwater pools in and among rocks, in hoofprints, vehicle ruts, and the like.

It is believed that there are at least four malaria vectors in the United States— *Anopheles freeborni* (West), *An. hermsi* (a relatively recently described species in the West), *An. punctipennis* (West), and *An. quadrimaculatus* (East) (*see* the discussion in Chap. 2, "Vector Competence," about these species in relation to malaria). *An. quadrimaculatus* is a complex of five identical looking species. Other species may also be involved in malaria transmission in the United States but are considered vectors of minor importance [10]. All four of the main vector species breed in permanent freshwater sites, such as ponds, pools, and rice fields, and are avid human biters. Accordingly, there is always the possibility of

**Fig. 3.9** Adult female *A. gambiae*

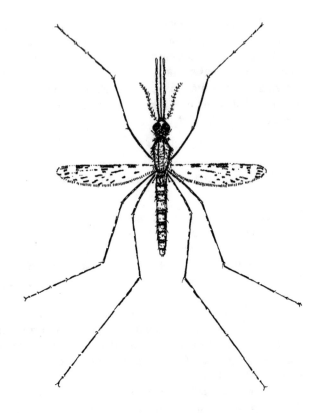

reintroduction of the malaria parasite into the United States and resumption of indigenously acquired cases.

### 3.2.5  Malaria Treatment and Control

#### 3.2.5.1  Prevention and Vector Control

Current malaria control programs are generally based on three primary interventions [11]:

1. Proper diagnosis and prompt treatment of human cases using artemisinin-based combination therapies (ACT) (*see* below)
2. Wide-scale distribution and use of insecticide-treated bed nets
3. Indoor residual spraying with insecticides to reduce vector populations, mostly using pyrethroids, although DDT may still be used as an indoor residual spray in some areas [12]

Certainly, protective clothing, insect repellents, and nets for sleeping may also reduce mosquito biting (see Appendix 3 for information on personal protection

**Fig. 3.10** Approximate geographic distribution of the *A. gambiae* complex

measures for mosquitoes). Bed nets impregnated with insecticide are even more effective. Since *Anopheles* mosquitoes bite at night, use of bed netting (properly employed) alone can significantly reduce risk of infection. Local vector control activities can also reduce malaria case numbers. This includes ultralow-volume insecticide fogging, usually by truck-mounted machines, to kill adult mosquitoes outdoors, larviciding to kill immature mosquitoes, and elimination of mosquito breeding habitats. Unfortunately, in countries with the worst malaria problems, financial resources are limited for mosquito control.

### 3.2.5.2  Malaria Vaccine

Ever since 1910, major efforts have been directed toward producing a malaria vaccine. There are obviously several points in the complex malaria life cycle where immunological interference with the multiplication of plasmodia could be attempted. Although many experimental vaccines have been developed and studied, no malaria

**Fig. 3.11** Adult female *A. darlingi*

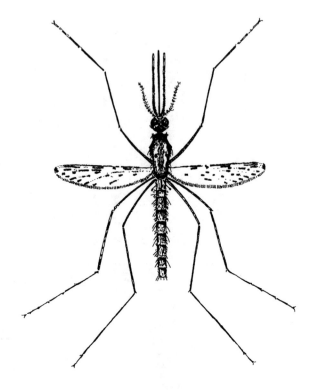

vaccine is widely available for use. However, the most advanced of the current candidates against *P. falciparum* malaria is the RTS,S vaccine which shows about 30% efficacy against malaria in young infants and children [13, 14]. There is progress with another promising vaccine called PfSPZ [5]. Development of a malaria vaccine has been stymied by several factors. For one thing, the persons at greatest risk of complications and death are young children, and most researchers expect that the initial immune responses elicited by a vaccine will be suboptimal. Second, even if a vaccine produces a vigorous humoral and cellular response, it does not necessarily provide sterile immunity. Even in naturally acquired infections, antibodies directed against the dominant antigen on the sporozoite surface do not prevent reinfection with sporozoites bearing the same dominant repetitive antigen. Third, for traditional vaccines, there are an inadequate number of adjuvants available for human use. For example, aluminum hydroxide is about the only adjuvant approved for human use. If other antigen–adjuvant combinations can be identified, which provide boosting with the reexposures that occur repetitively with natural reinfection under field conditions, then there is better hope for malaria control through vaccines.

### 3.2.5.3  Antimalarial Drugs

Drugs are primarily used for malaria control in two ways—*prevention* of clinical malaria (prophylaxis) and *treatment* of acute cases. Antimalarial drugs include chloroquine, primaquine, mefloquine, malarone, doxycycline, quinine (not used much

**Fig. 3.12** Approximate geographic distribution of *A. darlingi*

anymore), and artemisinin (usually in combination with other antimalarials). Because of increased parasite resistance to antimalarial drugs, treatment regimes have become quite complicated and vary tremendously by geographic region. In addition to the problem of resistance, serious side effects may occur with the use of some antimalarial products. The most effective current antimalarial treatment is artemisinin-based combination therapy (ACT) for countering the spread and intensity of *Plasmodium falciparum* resistance to chloroquine, sulfadoxine/pyrimethamine, and other malarial drugs [15]; however, there is documented resistance to ACT in parts of Southeast Asia which seems to be spreading [3]. Healthcare providers should contact their local or state health department, the CDC, or the preventive medicine department at a local medical school for the most up-to-date malaria treatment recommendations.

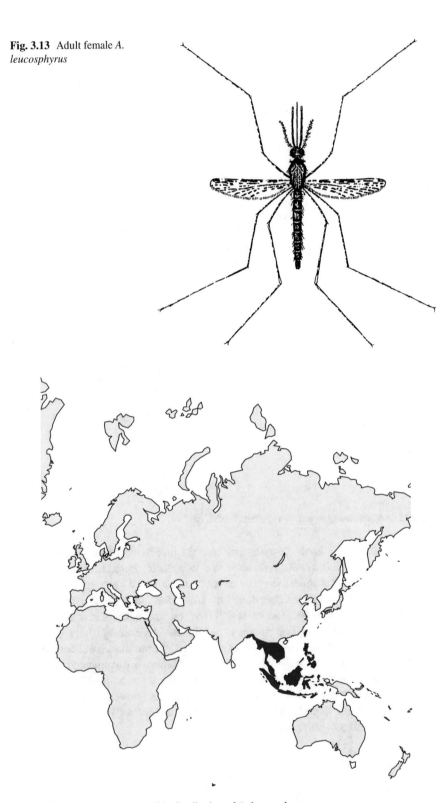

**Fig. 3.13** Adult female *A. leucosphyrus*

**Fig. 3.14** Approximate geographic distribution of *A. leucosphyrus*

**Table 3.1**  Characteristics of some encephalitis viruses in the United States

| Disease | Where it occurs | Mosquitoes | Mortality |
|---|---|---|---|
| St. Louis encephalitis (SLE) | Most of the United States, parts of Canada | *Culex quinquefasciatus, Cx. tarsalis, Cx. nigripalpus* | 3–20% |
| West Nile virus (WNV) | Africa, Asia, Europe, North America, parts of central and South America | Primarily *Culex* mosquitoes, especially *Cx. quinquefasciatus, Cx. pipiens,* and *Cx. tarsalis* | 4–12% |
| Eastern equine encephalitis (EEE) | Eastern and north Central United States, most common along Atlantic and gulf coasts | *Culiseta melanura* (enzootic) *Aedes sollicitans* and *Coquillettidia perturbans* (epizootic), others | 30–60% |
| Western equine encephalitis (WEE) | Western and Central United States, Canada | *Cx. tarsalis* | 2–5% |
| La Crosse (LAC) | Midwestern and southeastern United States | *Aedes triseriatus*, others | 1% |
| Venezuelan equine encephalitis (VEE) | Occasionally extreme southern United States (mostly central and South America) | *Psorophora columbiae* and *Aedes* spp. | 1% |

# 3.3   Mosquito-Transmitted Encephalitis Viruses

## 3.3.1   Introduction and General Comments

There are numerous mosquito-borne viruses in the world. In the United States, the most common ones are encephalitis viruses (Table 3.1). Generally, these viruses are zoonoses that circulate among small mammals or birds with various mosquitoes serving as vectors. Humans may become involved when conditions favor increased virus activity or geographic coverage. These outbreaks may be cyclical. For example, there is a 10-yr cycle of St. Louis encephalitis. There was an outbreak in New Jersey–Pennsylvania region in 1964 and another much larger outbreak in the Mississippi River Valley area in the mid-1970s [1]. West Nile virus appears to have a 5-yr cycle. Certainly not all cases of encephalitis are mosquito-caused (enteroviruses and other agents are often involved), but mosquito-borne encephalitis has the potential to become a serious cause of morbidity and mortality covering widespread geographic areas of the United States each year.

## 3.3.2   Eastern Equine Encephalitis (EEE)

### 3.3.2.1   The Disease

Of all the North American mosquito-borne encephalitis viruses, the one causing EEE is the worst. EEE is a severe disease of horses and humans having a mortality rate of 30–60%; there are also frequent neurological sequelae. Although some cases

**Mosquito-Borne Encephalitis: Test your Diagnostic Skills**

The following two cases of encephalitis are actual ones the author helped investigate while at the Mississippi Department of Health. See if you can figure out which encephalitis virus was involved in each case. I assure you, they are common ones. While your diagnostic decisions will be hampered by lack of specific laboratory findings, other clues—such as age of the patient, relative severity of the illness, clinical presentation, and kinds of mosquitoes collected—are revealed in the case histories.

**Case 1**. In late June 1999, a 6-month-old, previously healthy infant was brought to an emergency department (ED) with a several-hour history of fever, with a maximum temperature of 38.7 °C (101.6 °F). He was experiencing a focal seizure characterized by uncontrollable blinking of the left eye, twitching of the left side of the mouth, and random tongue movement. In the ED, seizures continued intermittently despite the administration of diazepam and lorazepam. Treatment was started with phenytoin, and the infant was admitted to the hospital.

Examination of cerebrospinal fluid (CSF) on admission showed 294 white blood cells (WBCs)/μL, with 47% polymorphonuclear leukocytes, 41% histiocytes, and 12% lymphocytes, and 3 red blood cells (RBCs)/μL. Protein and glucose levels were within normal limits. An admission CT scan was read as normal. Therapy with acyclovir was initiated because of the possibility of herpes encephalitis, and cefotaxime and vancomycin were also started to cover possible bacterial infection. The seizures stopped; the intravenous phenytoin was discontinued, and treatment with oral phenobarbital was started.

Focal seizures, which progressed to generalized tonic–clonic seizures, recurred on the fourth hospital day, although a therapeutic blood level of phenobarbital had been achieved. After another CT scan, which was read as within normal limits, the child was transferred to a different hospital.

On admission, the infant was noted to have continuous seizure-like movements of the chin and face. A lumbar puncture at this time revealed 307 WBCs/μL, with 33% polymorphonuclear leukocytes, 29% lymphocytes, and 38% histiocytes, 812 RBCs/μL, a protein level of 104 mg/dL, and a glucose level of 74 mg/dL. Treatment with acyclovir, cefotaxime, and vancomycin was continued. The patient was intubated because of excessive secretions and to avoid respiratory compromise. He was again treated with lorazepam, and phenytoin was restarted. Seizure activity ended. The patient continued to be intermittently febrile. He was extubated on the sixth continuous hospital day.

Results of polymerase chain reaction studies for herpesvirus from admission samples were negative. In addition, admission blood, urine, and CSF culture results were negative, and on the ninth hospital day, cefotaxime and vancomycin were discontinued. The patient's maximum temperature on that day was 37.8 °C (100.1 °F), and he was becoming more alert and playful.

(continued)

He subsequently became and remained afebrile, without seizure activity, so he was transferred back to the original hospital on the 12th hospital day for completion of 21 d of intravenous acyclovir. On day 21, he was discharged home on a regimen of oral phenytoin. He had some residual left-sided weakness requiring several weeks of physical therapy.

A sample of CSF was sent to a reference laboratory for testing for antibodies to various encephalitis agents, including herpesviruses and arboviruses. Results showed an indirect fluorescent antibody titer of 1:8 to a mosquito-borne encephalitis virus. Serum was sent to the CDC for confirmation, which showed the presence of IgM antibody to that same virus. Mosquito collections at the patient's home revealed *Culex restuans*, *Culex salinarius*, *Aedes albopictus*, *Ae. triseriatus*, *Anopheles crucians*, and *An. quadrimaculatus*. Based on this description, what disease would you have diagnosed?

**Case 2**. Fever (temperature of 39.4 °C [103 °F]) and diarrhea developed in an 11-year-old Native American boy on July 31. Gastroenteritis was reportedly "going around" in the community at the time. He was taken to a local ED and given symptomatic treatment. His condition improved somewhat until the night before admission, when he complained of headache, stomachache, and decreased appetite. He went to bed early, which was unusual for him. The next morning he went with his family to a scheduled ophthalmologic examination and slept during most of the 1-h drive. He was drowsy and nauseated on arrival at the clinic; he turned pale and grand mal seizure activity began. He was taken to the ED and given a loading dose of phenytoin, and then he was transferred to the admitting hospital.

On admission, he was responsive but lethargic. His admission temperature was 39°C (102.2°F). His admission laboratory findings included a WBC count of 19,500/μL, with 59% neutrophils, 19% band forms, and 16% lymphocytes. The hematocrit was 34.4%, and the serum glucose level was 184 mg/dL. CSF examination revealed a WBC count of 980/μL, with 91% neutrophils, no organisms on Gram stain, a negative latex agglutination test, a protein level of 68 mg/dL, and a glucose level of 105 mg/dL. Additional blood, CSF, and stool cultures were obtained.

The patient was given cefotaxime and phenytoin. He remained febrile, with temperatures up to 40.6°C (105°F), but became more responsive and became ambulatory by the second day after admission. At 2 pm on August 5, the patient experienced another seizure, with eye deviation to the right and head turning to the right. A CT scan showed enhancement of the cisterna but only mildly increased intracranial pressure. Respirations became irregular, and the patient was electively intubated and hyperventilated. Treatment with streptomycin, pyrazinamide, and isoniazid was started for possible CNS herpesvirus infection. His condition deteriorated over the next 24 h until he showed no evidence of brain stem function. A lumbar puncture was performed

(continued)

for viral studies, since none of his previous cultures were growing. He was taken off the ventilator the evening of August 6. At autopsy, the patient's meninges were relatively clear, but cerebral edema was present.

Confirmation of infection with one of the mosquito-borne encephalitis viruses was made by the CDC facility in Ft. Collins, CO; two separate serum samples indicated a fourfold rise in hemagglutination inhibition antibody to the virus, and enzyme-linked immunosorbent assay showed the presence of specific IgM. Mosquito species collected around the home included *Aedes albopictus, Anopheles crucians,* and *Coquillettidia perturbans.* Which arboviral disease did this child have?

Answers.

**Case 1**. La Crosse encephalitis (LAC). This infection occurs mostly in children, and seizures are quite commonly the presenting symptom, occurring in about 50% of clinical cases. Most cases are mild; the mortality rate is only about 1%. Another clue to the virus' identity in this case is the mosquito collection data. The primary vector mosquito for LAC is *Ae. triseriatus.*

**Case 2**. Eastern equine encephalitis (EEE). This is a devastating disease with a mortality rate of about 50%. As exemplified in this case, EEE is especially severe in children. Within just a few days, this boy was dead. Again, in this case, the mosquito data provide an additional clue: the primary inland vector of EEE is *Cq. perturbans.*

may be asymptomatic, most are characterized by acute onset of headache, high fever, meningeal signs, stupor, disorientation, coma, spasticity, tremors, and convulsions [1, 16]. The disease is especially severe in children. I helped investigate a fatal case in an 11-yr-old boy who exhibited headache, anorexia, and excessive sleepiness on the day of hospital admission (see textbox above). Later, he developed nausea and fever and started grand mal seizure activity. At day 3, respirations became irregular, and he eventually showed no signs of brain stem function [17].

### 3.3.2.2   Ecology of EEE

EEE occurs in the central and north central United States and especially along the Atlantic and Gulf Coasts (cases can occasionally occur several 100 miles inland; *see* Fig. 3.15) [18]. More recently, it has been expanding into northern New England. The

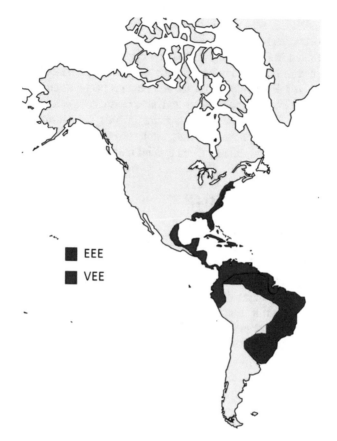

**Fig. 3.15** Approximate geographic distribution of eastern equine and Venezuelan equine encephalitis

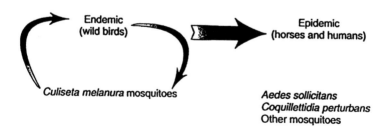

**Fig. 3.16**  Life cycle of EEE. (Provided with permission by Infections in Medicine 1996;13:671)

appearance of EEE is seasonal; in the southernmost areas of the virus range, human cases may occur year-round but are concentrated between May and August. EEE virus is sustained in freshwater swamps in a cycle involving birds and mosquitoes with the main enzootic vector being *Culiseta melanura*, which rarely bites humans or horses (Fig. 3.16). Epidemics in horses and humans occur when prevalence of the virus in bird populations becomes high and other mosquito species become involved. These secondary or epizootic vector species include the salt-marsh mosquitoes, *Aedes sollicitans* and *Ae. taeniorhynchus*, on the coast, and the freshwater mosquito, *Coquillettidia perturbans*, inland. Other species may also be involved [19].

### 3.3.2.3   Differential Diagnosis of EEE

Differentiation must be made from other encephalitides (postvaccinal or postinfection); tick-borne encephalitis; rabies, non-paralytic polio; mumps meningoencephalitis; aseptic meningitis from enteroviruses; herpes encephalitis; various bacterial, mycoplasmal, protozoal, leptospiral, and mycotic meningitides or encephalitides; and others [16]. Any cases of encephalitis in mid- to late summer should be suspected. Specific identification is usually made by finding specific IgM antibody in acute serum or cerebrospinal fluid (CSF) or antibody rises (usually HI test) between early and late serum samples.

### 3.3.2.4   Control of EEE

Since there is no specific treatment available for EEE or any other mosquito-borne encephalitis, control of the disease rests entirely on either preventing transmission to humans or interrupting the virus cycle in nature. Preventing transmission involves personal protective measures against mosquito biting, such as avoiding outdoor activity after dark, wearing long sleeves, and judicious use of repellents such as those containing the active ingredients DEET, or picaridin, or oil of lemon eucalyptus (see Appendix 3). Interrupting the EEE virus cycle in nature involves spraying the area (by ground equipment or by airplane) for adult mosquito control, as well as environmental sanitation efforts in affected communities to eliminate mosquito breeding sites.

Sometimes an environmental survey of the area where cases occur can lead to further prevention recommendations. For example, in the fatal case of EEE mentioned above [17], the patient lived in a house without window screens. This likely led to increased exposure to mosquitoes (and thus biting)—a risk factor for any mosquito-borne disease. This was of interest, since we take for granted the fact that basic sanitation and public health measures, such as screen wire windows, are implemented. Also, a mosquito survey at the patient's house revealed numerous prime *Coquillettidia perturbans* (the suspected mosquito vector in this case) breeding sites. In addition, *Cq. perturbans* were collected by CDC light traps in the community at the time of survey. Accordingly, control efforts were directed toward that particular mosquito species, thus averting new cases.

### 3.3.3 St. Louis Encephalitis (SLE)

#### 3.3.3.1 The Disease

Outbreaks of SLE are sporadic and somewhat cyclical. For example, in 1933, there were 1095 cases of SLE with more than 200 deaths [20]. About 40 yr. later, another major outbreak occurred in the Mississippi Valley region with over 2000 cases [20]. SLE is worse in elderly patients; young people often have no clinical symptoms or only mild, influenza-like symptoms (note: this is just the opposite of EEE). There is usually abrupt onset of fever, headache, and malaise. Physical exam may only reveal elevated temperature and perhaps dehydration. Over a period of several days to a week, other signs of central nervous system (CNS) infection may develop, such as stiff neck, disorientation, tremulousness, unsteadiness, confusion, and even coma. The mortality rate is 3–20%. One elderly patient I interviewed spoke of an extreme fatigue, forcing him to bed, persisting for weeks after the infection. The clinical features of SLE are not specific, so the illness must be differentiated from other etiologies, such as bacterial, other viral, mycobacterial, fungal, rickettsial, toxic, cerebrovascular, and neoplastic diseases [21]. Time of the year may be a clue to recognizing SLE, since most cases occur in mid- to late summer. Clinical laboratory results are generally not distinctive. CSF may show a preponderance of polymorphonuclear cells if obtained early in the illness; a shift toward lymphocytic pleocytosis is the rule [22]. Protein in the CSF may be slightly elevated above normal during the first and second weeks of illness. Definitive diagnosis is usually made by detection of specific IgM in acute serum or CSF.

#### 3.3.3.2 Ecology of SLE

In the United States, SLE virus circulates in nature among birds, being transmitted by bird-biting mosquitoes in the genus *Culex* (Fig. 3.17). Susceptible birds become viremic, and infect new mosquitoes feeding on them, which then, in turn, infect new birds. This cycle continues year to year, with no apparent effect on the birds or mosquitoes. For various reasons—climatic factors or numbers of mosquitoes or birds—the virus level in nature is amplified to the point where aberrant hosts (humans) become infected. Interestingly, times of drought may actually enhance SLE transmission by concentrating vector mosquitoes and bird hosts [23]. There are four main mosquito vectors of SLE in the United States: *Culex tarsalis* in the West, *Cx. pipiens* (northern) and *Cx. quinquefasciatus* (southern) in the East, and *Cx. nigripalpus* in Florida (Fig. 3.18). *Cx. tarsalis* (Fig. 3.19) is primarily a rural species that breeds in both polluted and clear water in ground pools, grassy ditches,

**Enzootic**

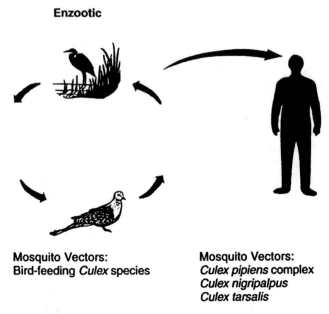

| | |
|---|---|
| **Mosquito Vectors:** | **Mosquito Vectors:** |
| Bird-feeding *Culex* species | *Culex pipiens* complex |
| | *Culex nigripalpus* |
| | *Culex tarsalis* |

**Fig. 3.17** Life cycle of St. Louis encephalitis virus. (Provided with permission by Infections in Medicine 1996;13:751)

and artificial containers. In arid regions, it is frequently found in canals and irrigation ditches. The *Cx. pipiens* complex (including northern and southern forms) breeds in ditches, storm sewer catch basins, cesspools, polluted water, and artificial containers around homes, such as cans and old tires. *Cx. nigripalpus* breeds in shallow rainwater pools, semipermanent ponds, and artificial containers.

### 3.3.3.3   Control of SLE

Control of SLE is basically the same as that of EEE (*see* Sect. 3.2.4); except the targeted mosquito vectors are different. Many healthcare workers fail to realize the specificity of viruses and their hosts and vector mosquitoes. All of these factors (animal hosts, vector mosquitoes, and so forth) are different for the particular virus involved. Rarely do generalized recommendations or control schemes work. For example, once I was investigating an EEE case in an adult male patient who lived in a rural community. The health department field investigator accompanying me proceeded to tell the local people about cleaning up around their homes—removing cans, tires, and so on—where container-breeding species could live. Since the primary inland vector of EEE in the area was a species that lived in marshes containing emergent vegetation (like cattails), this health department worker was spreading misinformation and, more importantly, was not in any way helping to prevent new

**Fig. 3.18** Approximate geographic distribution of St. Louis encephalitis

cases. Just as is the case with microbes in which specific control depends on the species and behavior of the organism, so it is with mosquitoes. Control measures must be targeted toward the specific vector species involved.

## 3.3.4   West Nile Encephalitis (WNV)

West Nile virus (WNV) was first detected in the Western Hemisphere in 1999 in New York City [24, 25]. This outbreak of mosquito-borne encephalitis was originally identified as St. Louis encephalitis (SLE) because the two viruses are closely related and cross-reactions occur with some serological lab tests. Over the next 5 yrs, WNV spread across the continental United States as well as north into Canada and southward into the Caribbean Islands and Latin America. In addition to being antigenically related, WNV and SLE have similar clinical profiles, life cycles, and

**Fig. 3.19** *Culex tarsalis*, a principal vector of St. Louis encephalitis in the Western United States. (From Carpenter and LaCasse, Mosquitoes of North America, University of California Press, 1955, now in public domain)

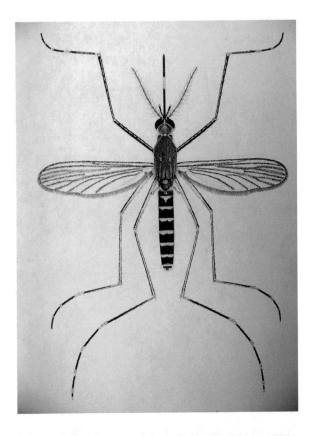

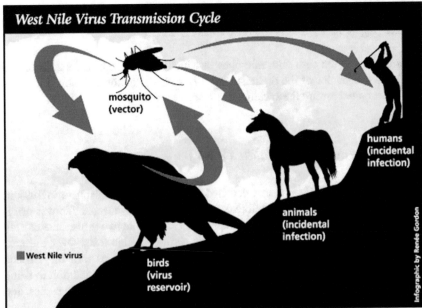

**Fig. 3.20** West Nile virus life cycle. (US FDA figure)

mosquito vectors (Fig. 3.20). WNV has been associated with significant human morbidity and mortality in the United States since its recognition; over 23,000 cases of fever or neuroinvasive disease have been reported to date to the Centers for Disease Control. As far as severity of the disease, WNV is no more dangerous than SLE (one of our "native" encephalitis viruses).

Approximately 80% of all WNV infections are asymptomatic and 20% cause West Nile fever, and less than 1% cause West Nile neuroinvasive disease [26]. Like SLE, WNV is more dangerous to older patients. Interestingly, of the first five patients in New York City admitted to hospitals, four had severe muscle weakness and respiratory difficulty, a finding atypical for encephalitis [27]. Also, GI complaints such as nausea, vomiting, or diarrhea occurred in four of five patients [27]. Much is yet to be learned about the ecology of WNV in the United States, but we do know the virus causes a bird disease and is transmitted by mosquitoes in the genus *Culex*. House sparrows and robins have been found to be among the best amplifying hosts in nature, producing highest viremias for the longest period of time. Although the virus has been isolated from at least 64 US mosquito species, the main vectors are believed to be *Culex pipiens*, *Cx. quinquefasciatus*, *Cx. salinarius*, *Cx. restuans*, and *Cx. tarsalis* [25, 28, 29].

### 3.3.4.1 Control of West Nile Encephalitis

Control of WNV is basically the same as that of SLE, which primarily involves finding and eliminating (or treating) breeding sites for *Culex quinquefasciatus*, *Cx. pipiens*, and *Cx. tarsalis*. In addition, ULV adulticiding for mosquitoes with truck-mounted or airplane-mounted sprayers can be useful in reducing populations of vector mosquitoes. Personal protection measures include wearing long sleeves and long pants when outdoors, proper screening and netting, and using of insect repellents (Appendix 3).

## 3.3.5 Other Mosquito-Borne Viruses

### 3.3.5.1 Western Equine Encephalitis (WEE)

WEE, transmitted mainly by *Cx. tarsalis*, occurs in the western and central United States, parts of Canada, and parts of South America and has erupted in several large outbreaks in the past (Fig. 3.21). There were large epidemics in the north central United States in 1941 and in the central valley of California in 1952. The 1941 outbreak involved 3000 cases [1]. Between 1964 and 1997, 639 human WEE cases were reported to CDC, for a national average of 19 cases/yr [30]. The incidence of WEE has been declining over the past few decades. WEE is generally less severe than EEE and SLE with a mortality rate of only 2–5%.

**Fig. 3.21** Approximate geographic distribution of western equine encephalitis

### 3.3.5.2    La Crosse Encephalitis (LAC)

La Crosse encephalitis (LAC) has historically affected children in the midwestern states of Ohio, Indiana, Minnesota, and Wisconsin (Fig. 3.22); however there have been large outbreaks in other states such as West Virginia. Serological and epidemiological studies in North Carolina, Georgia, Tennessee, Mississippi, Virginia, and Florida have indicated that LAC occurs and is increasing in those states also, but not to the extent that it currently occurs in the Midwestern United States. The mortality rate of LAC is <1%, but seizure disorder may follow LAC infection. During the 33-yr period (1964–1997), 2497 LAC encephalitis cases were reported to CDC, at an average of 73 cases/yr [30]. Interestingly, LAC virus may be transferred from adult female *Ae. triseriatus* to her offspring through ovarial contamination.

**Fig. 3.22** Approximate geographic distribution of La Crosse encephalitis. (Adapted from WHO publication WHO/VBC/89.967)

Amplification of the virus takes place in nature through an *Ae. triseriatus*-wild vertebrate cycle.

### 3.3.5.3   Venezuelan Equine Encephalitis (VEE)

VEE is relatively mild and rarely affects the CNS but will be included here as an encephalitis. The virus is transmitted by many mosquito species, particularly those in the genus *Psorophora*. The mosquito, *Ae. taeniorhynchus*, was found to be an important vector during the most recent outbreak in Columbia and Venezuela [31]. VEE is endemic in Mexico and Central and South America; epidemics occasionally reach the southern United States (Fig. 3.15). Although the mortality rate is generally <1%, significant morbidity is produced by this virus.

**Fig. 3.23**  Approximate geographic distribution of Japanese encephalitis

In an outbreak in Venezuela in 1962–1964, there were more than 23,000 reported human cases with 156 deaths [1]. In 1971, an outbreak of VEE in Mexico extended into Texas resulting in 84 human cases [32]. A more recent outbreak in Colombia and Venezuela (1995) resulted in at least 75,000 human cases [31].

### 3.3.5.4   Japanese Encephalitis (JE)

JE does not occur in the United States but is the principal cause of epidemic viral encephalitis in the world with almost 70,000 clinical cases occurring annually; some 15,000 die from the illness each year [33, 34]. JE is highly virulent. Approximately 25% of cases are rapidly fatal, 50% lead to neuropsychiatric sequelae, and only 25% fully recover [35]. As many as 85% of children with JE have seizures. The virus is transmitted by several *Culex* mosquitoes, but especially *Cx. tritaeniorhynchus*. JE epidemics have, at times, been widespread and quite severe. Historically, JE has been focused in the northern areas of countries in Southeast Asia, East Asia, and midsouthern Asia, especially China and Vietnam

(Fig. 3.23). Recently, there has been a steady westward extension of reported epidemic activity into northern India, Nepal, and Sri Lanka. JE has the potential for introduction and establishment in North America, especially via international travel and smuggling of animals and legal exotic pets [36]. There is an effective vaccine for JE, and substantial progress has been made in implementing immunization programs in countries with JE virus transmission risk.

### 3.3.5.5 Chikungunya

Chikungunya (CHIK) is a mosquito-transmitted Alphavirus which is not usually fatal but can cause severe fevers, headaches, fatigue, nausea, and muscle and joint pains [37, 38]. There is often excruciatingly painful swelling of the joints in fingers, wrists, back, and ankles. The virus was first isolated during a 1952 epidemic in Tanzania, and the word Chikungunya comes from Swahili, meaning "that which bends up," referring to the position patients assume when suffering severe joint pains [38]. The geographic distribution of CHIK has historically included most of sub-Saharan Africa, India, Southeast Asia, Indonesia, and the Philippines, although the disease is increasing both in incidence and geographic range. There were at least 300,000 cases on Reunion Island in the Indian Ocean during 2005–2006. India suffered an explosive outbreak in 2006 with more than 1.25 million cases. CHIK was found in Italy in 2007 [37, 39]. One of the main mosquito vectors of CHIK is the Asian tiger mosquito, *Aedes albopictus*, which is extremely abundant in the southern United States, raising fears of widespread outbreaks should local mosquitoes become infected [39]. In 2013–2014, there were thousands of cases reported in the Caribbean and Central and South America [40]. In 2015, there were 896 cases of travel-related CHIK in the United States and one locally acquired case [41].

### 3.3.5.6 Zika Virus

Zika virus (ZIKV) is primarily a mosquito-borne human disease with only limited clinical severity but which can still be serious due to devastating birth defects (microcephaly) in pregnant females. The virus is a positive-sense single-stranded RNA virus in the family Flaviviridae, which includes several other viruses of clinical importance. ZIKV is closely related to Spondweni virus, the only other member of its clade. Phylogenetic analysis shows that ZIKV can be classified into distinct African and Asian lineages; both emerged from East Africa during the late 1800s or early 1900s [42]. In humans, the incubation period is approximately 3–12 days. Infection is asymptomatic in 80% of cases, making reporting and (necessary) public health interventions difficult. People of all ages are susceptible to ZIKV (4 days– 76 yrs), with a slight preponderance of cases in females [42, 43]. When symptoms occur, they are typically mild, self-limiting, and nonspecific. Commonly reported symptoms include rash, fever, arthralgia, myalgia, fatigue, headache, and conjunctivitis [44]. Rash, a prominent feature, is maculopapular and pruritic and usually

**Fig. 3.24** Microcephaly
caused by Zika virus.
(CDC figure)

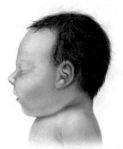

Baby with Typical Head Size

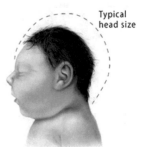

Baby with Microcephaly

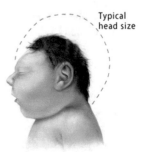

Baby with
Severe Microcephaly

begins proximally and then spreading to the extremities with spontaneous resolution within 1–4 days of onset [42]. Fever is usually low grade (37–38°C). All symptoms resolve within 2 weeks and reports of longer persistence are rare. More severe clinical sequelae have increasingly been associated with Zika virus, including Guillain-Barré syndrome and microcephaly (Fig. 3.24) [43]. During the 2015 outbreak in Brazil, reports of infants born with microcephaly greatly increased (>3800 cases; 20 cases/10,000 live births vs 0.5/10,000 live births in previous years) [42]. During 2015–2017, there were millions of cases of ZIKV throughout the Americas and approximately 275 locally acquired cases in the United States [44, 45]. All cases in the United States have thus far been associated with *Ae. aegypti* mosquitoes and not *Ae. albopictus.*

Zika virus, like other flaviviruses, is transmitted by mosquitoes, primarily genus *Aedes* (*Stegomyia*). Several *Aedes* spp. have been implicated, including *Ae. aegypti, Ae. africanus, Ae. hensilli, and Ae. albopictus.* In the United States and the Caribbean, Zika is primarily transmitted by the yellow fever mosquito, *Aedes aegypti,* and (potentially) the Asian tiger mosquito, *Ae. albopictus.* Both species are very similar in appearance and habits. *Aedes aegypti* is a small black species with prominent white bands on its legs and a silver-white lyre-shaped figure on the upper side of its thorax. It breeds in artificial containers around buildings such as tires, cans, jars, flowerpots, and gutters and usually bites during the morning or late afternoon. They may readily enter houses and seem to prefer human blood meals (as opposed to animals), biting principally around the ankles or back of the neck. Interestingly, in many places in the United States where *Ae. albopictus* was accidentally introduced, *Ae. aegypti* has virtually disappeared, apparently being displaced.

*Aedes albopictus* is widely distributed in the Asian region, the Hawaiian Islands, parts of Europe, and much of the Americas, including the southern United States, where it was accidentally introduced in 1986 [46, 47]. This species is similar in appearance to *Ae. aegypti* having a black body and silver-white markings, with the major difference between the two being that *Ae. albopictus* has a single, silver-white stripe down the center of the dorsum of the thorax (instead of the lyre-shaped marking) (Fig. 3.25). They also breed in artificial containers such as cans, gutters, jars, tires, flowerpots, etc., and seem especially fond of discarded tires. This is an aggressive daytime-biting mosquito, often landing and biting immediately.

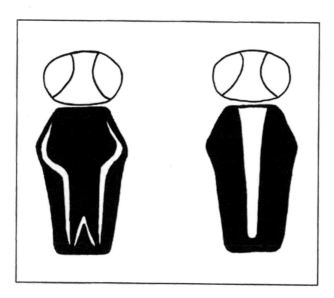

**Fig. 3.25** Markings on thorax of mosquito vectors of dengue in the United States *A. albopictus* on the right. (provided with permission by Infections in Medicine 1996;13:933)

## 3.4   Dengue Fever

### *3.4.1   Introduction*

Dengue (DEN) is a serious mosquito-borne human disease occurring in the tropical countries of Asia and Africa, as well as in the Caribbean area, and Central and South America. The virus, a flavivirus related to yellow fever virus, has four serotypes (Den 1, Den 2, Den 3, and Den 4) and is transmitted to people primarily by the mosquitoes, *Ae. aegypti* and *Ae. albopictus*. As far as is known, humans are the main vertebrate reservoir of the virus, although there may be a monkey-mosquito cycle in some areas. The disease is characterized by fever, headache, retro-orbital pain, and intense aching; for this reason it is sometimes referred to as "breakbone fever." Occasionally, a more severe form of the disease occurs, dengue hemorrhagic fever/dengue shock syndrome (DHF/DSS), which may result in a hemorrhagic shock syndrome with a fatal outcome.

Dengue is not some small, insignificant disease entity—there are approximately 100 million cases each year [48, 49]. One research study using new modeling techniques raised this estimate of dengue cases to 390 million per year [50]. Currently, it is not widely endemic in the United States but is literally "knocking at the door" (Fig. 3.26). Hundreds of cases occurred in the summer of 1995 along the Texas–Mexico border, especially in the Reynosa area. There was a fairly large outbreak of dengue in the Florida Keys during 2009–2010 [51]. There is always the possibility of a widespread dengue epidemic in the United States since there is an abundance

**Fig. 3.26** Approximate worldwide distribution of dengue

of the mosquito vectors in the south central United States. Also, with the thousands of people returning home from cruises to the Caribbean each month (especially during the summer), there is good chance of infected persons returning and infecting local mosquitoes with the virus.

## 3.4.2  Spread of the Virus

Dengue virus is transmitted by the bite of an infected female mosquito. Mosquitoes may become infected by feeding on viremic patients, generally only from the day before to the end of the febrile period. Usually, they will not feed again for 3–5 d, depending on temperature. It is in this second (or third, rarely) feeding when a susceptible person is inoculated with the virus. The adult lifespan of dengue vector mosquitoes is generally very short (few days), although some may survive 14 d or longer. Accordingly, it is amazing that dengue virus transmission occurs at all. However, mosquito populations are so great that even though most females die before feeding a second time, enough individuals survive long enough to keep virus transmission going.

Ae. aegypti and Ae. albopictus are the mosquito vectors of dengue in the Western Hemisphere. They are somewhat similar in appearance, although markings on their thorax ("back" of mosquito, where wings are attached) are different (Figs. 3.25, 3.27, and 3.28). They both are similar in habits and breeding sites, feeding in the daytime (mostly early and late) and breeding in artificial containers around the home. Prime sites include paint cans, old tires, vases and jars, clogged rain gutters, pet watering dishes, etc. *Aedes albopictus*, known as the Asian tiger mosquito, was accidentally introduced into the United States from Japan in 1985 in the Houston, Texas area. Since then, it has rapidly spread over much of the central and southern United States, often replacing the native Ae. aegypti [52] (Figs. 3.29 and 3.30). Today, the Asian tiger mosquito is the primary pest mosquito in many towns and cities, being extremely difficult to control by standard mosquito spraying (trucks or airplane) because of its close proximity to houses and daytime feeding habits.

## 3.4.3  Clinical and Laboratory Characteristics

Clinically, dengue may be difficult to differentiate from leptospirosis, malaria, typhoid, measles, yellow fever, or chikungunya [49]. After an incubation period of 5–8 d, there is sudden onset of fever, severe headache, retro-orbital pain, myalgia, and arthralgia. There may also be GI disturbances, mottling of the skin, and rash. Severe hemorrhagic manifestations (DHF/DSS) may occur, especially in children, and is thought to be a result of, among other things, a sequential infection by more than one dengue virus serotype (*see* next section) and/or variations in viral virulence. DHF/DSS is characterized by fever, excessive capillary permeability, hypovolemia, and abnormal blood clotting mechanisms. Frequently reported hemorrhagic

**Fig. 3.27** Adult female *A. aegypti* mosquito. (From Carpenter and LaCasse, Mosquitoes of North America, University of California Press, 1955, now in public domain)

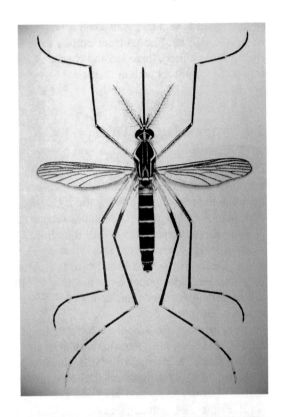

**Fig. 3.28** Adult *A. albopictus*. (CDC photo by James Gathany)

signs are scattered petechiae, a positive tourniquet test, easy bruisability, and less frequently, epistaxis, bleeding at venipuncture sites, a petechial rash, and gum bleeding. Although historically DHF/DSS has been mostly reported from Southeast Asia, it is increasingly being seen in the Western Hemisphere [53]. For example, there was an epidemic of DHF/DSS in Cuba in 1981 with >10,000 cases of severe hemorrhagic fever and 158 deaths [54]. In August 2005, health authorities in Tamaulipas, Mexico, reported 223 cases of DHF [53].

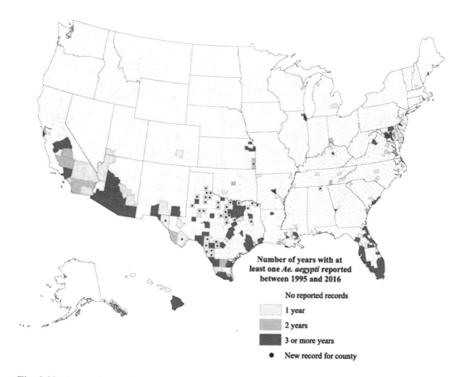

**Fig. 3.29** Approximate distribution of *Aedes aegypti* in the United States. (From [52]. Published by Oxford University Press on behalf of Entomological Society of America 2017. This work is written by US Government employees and is in the public domain in the United States)

### 3.4.3.1   Risk Factors for DHF/DSS

The following are various host and virus factors believed to convert a benign and self-limiting disease, dengue, into the severe syndrome, DHF/DSS. This list comes from Halstead [55].

1. Infection parity: An overriding risk factor for DHF/DSS in individuals >1 yr old is history of one prior dengue infection.
2. Passively acquired dengue antibody: Antibodies to dengue acquired transplacentally place infants at high risk for DHF/DSS during a first dengue infection during the first year of life.
3. Enhancing antibodies: Dengue virus infection-enhancing antibody activity in undiluted serum is strongly correlated with DHF/DSS in individuals who experience a subsequent secondary dengue infection.
4. Absence of protective antibodies: Low levels of cross-reactive neutralizing antibody protect, but DHF/DSS occurs in their absence.
5. Viral strain: DHF/DSS is associated with secondary infections with dengue viruses of Asian origin.
6. Age: DHF/DSS is usually associated with children.

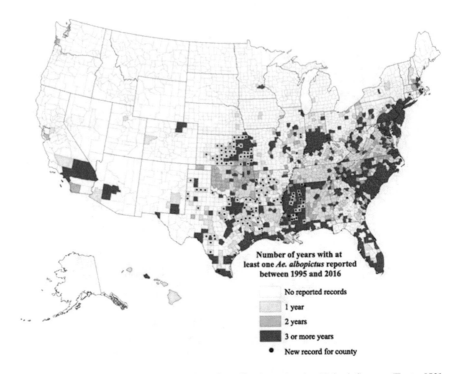

**Fig. 3.30** Approximate distribution of *Aedes albopictus* in the United States. (From [52]. Published by Oxford University Press on behalf of Entomological Society of America 2017. This work is written by US Government employees and is in the public domain in the United States)

7. Sex: Shock cases and deaths occur more frequently in female than male children.
8. Race: During the 1981 Cuban epidemic, Blacks had lower hospitalization rates for DHF/DSS than Asians or Whites.
9. Nutritional status: Moderate to severe protein-calorie malnutrition reduces risk of DHF/DSS in dengue-infected children.
10. Preceding host conditions: Menstrual periods and peptic ulcers are risk factors for the severe bleeding in adults, which occurs during some dengue infections.

### 3.4.4   Treatment, Prevention, and Control

There is no specific treatment for dengue. However, the hypovolemic shock resulting from DHF may require several specific interventions (*see* an appropriate clinical text for current guidelines). Prevention and control of dengue in a community depend on personal protection measures against mosquito biting (repellents, screening, long sleeves, and so forth; *see* Appendix 3) and reducing populations of the two vector mosquitoes. Both of these require public education campaigns. Since the traditional ultralow-volume insecticide sprays are mostly ineffective against these species, elimination of larval breeding sources is needed. This requires convincing the public and property owners of the need for such activity. Special clean-up days may need to be proclaimed by government officials to promote elimination of breeding sites around homes (remember, this species is not ordinarily found deep in the woods or swamps). If necessary, special teams of health department or volunteer evaluators may be formed to walk through every neighborhood, inspecting premises, dumping out water-filled containers ("tip and "toss"), and possibly treating other breeding sites with pesticide.

### 3.4.4.1 Dengue Vaccine

There are several candidate dengue vaccines currently making their way through clinical trials [56]. The only one currently registered (or licensed) is Dengvaxia® by Sanofi Pasteur, which has shown modest efficacy in preventing dengue infection. In trials, Dengvaxia® reduced dengue disease by 56.5% but provided only limited protection against the Den 2 serotype [57, 58]. Further, it did not have the hope for high and balanced efficacy over all age groups [58]. Making matters worse, there have been safety concerns with Dengvaxia®, and the company now says its use will be dramatically curtailed [59].

## 3.5 Yellow Fever (YF)

### 3.5.1 Introduction and Medical Significance

YF is probably the most lethal of all the arboviruses and has had a devastating effect on human social development. It consists of a sylvatic form—that can occasionally spread to humans—circulating among forest monkeys in Africa and Central and South America and an urban form transmitted by the peridomestic mosquito, *Aedes aegypti*, without any involvement of monkeys [60] (Fig. 3.31). The illness is characterized by a dengue-like syndrome with sudden onset of fever, chills, headache, backache, myalgia, a flush face, prostration, nausea, and vomiting. Jaundice is usually moderate early in the disease and intensified later. In one study, all patients with confirmed YF had jaundice, and 50% hemorrhaged from the nose and gums [61]. After the initial clinical syndrome, there may be a brief remission lasting from several hours to a day. Then the patient may enter a period of severe disease, including intoxication, renewed fever, hemorrhages, hematemesis (the classic sign of YF—"black vomit"), albuminuria, and oliguria. Published sensational accounts often mention profuse vomiting of black material, collapse, and sudden death.

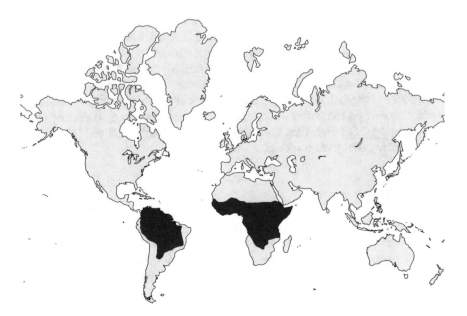

**Fig. 3.31**  Approximate geographic distribution of yellow fever

YF is caused by a flavivirus that is transmitted to humans by the bite of infected mosquitoes. Even though an effective vaccine is available, there is still a significant disease burden in Africa and South America from YF. Newly derived estimates of case numbers are 180,000 people sickened by YF each year and 78,000 deaths [62]. The case fatality rate among indigenous people of endemic areas is <5% but may exceed 50% in nonindigenous groups and during epidemics. Although it can occur from Mexico to Argentina, most cases in the Americas occur in northern South America and the Amazon basin, including the Colombian llanos (extensive plains) and eastern regions of Peru and Bolivia. There has been a recent outbreak in Brazil. In Africa, the endemic zone is the humid tropics roughly between about 16°N and 10°S latitudes. The most recent outbreak there has been in Angola [63]. Interestingly, YF does not occur in India or the densely populated countries of Southeast Asia, even though plentiful vector mosquitoes and monkeys occur there. Historically, YF epidemics have hit European seaports and many American towns and cities with devastating results (*see* Sect. 3.5.2). YF must never be ignored lest it be reintroduced into nonendemic areas via fast-paced air travel. The vector mosquitoes are ever present; all that is needed is the introduction of the etiologic agent. In fact, there was an imported case of YF in a 45-yr-old man who returned to Tennessee from a 9-d trip to Brazil [64]. Reportedly, he had visited the jungles of Brazil along the Rio Negro and Amazon Rivers and had not been previously vaccinated. The man died 6 d after hospital admission and 10 d after his first symptoms appeared.

## 3.5.2 Brief History of Yellow Fever

YF probably originated in Africa and was transported via shipping to port cities in the New World, where from 1668 to 1893, it erupted in 135 major epidemics, leaving economic shambles, human panic and fear, and widespread death [65]. New Orleans, Memphis, and Philadelphia were among the hardest hit [66]. In 1793, one of every ten Philadelphians died. In New Orleans during a major outbreak in 1853, there were 29,000 cases with over 8000 deaths. One of the worst American epidemics occurred in 1878 involving 132 towns, 75,000 cases, and 16,000 deaths. During that epidemic, Memphis was totally devastated [66, 67]. Panic ensued. People with financial resources fled, leaving behind only the poor, sick, or dying. The city was considered dead. Below is an eyewitness account given by William T. Ramsey of Washington, who came to Memphis with a corps of Howard Association nurses [67]:

- Memphis is a city of horrors. The poor whites and Negroes from 150 miles around Memphis have flocked into the city looking for food. Hundreds of them prowl around the streets with hardly any clothes on. They break into the vacant houses whenever they want. The stench of Memphis sickened me before I got within 5 miles of the city. No words can describe the filth I saw, the rotten wooden pavements, the dead animals, putrefying human bodies, and the half-buried dead combining to make the atmosphere something fearful. I took 30 grains of quinine and 120 drops of tincture of iron every day and wore a thick veil soaked with carbolic acid over my face. Many of the nurses, both men and women, smoke cigars constantly while attending patients to ward off the stench. In the Peabody Hotel where I stayed, pans of sulfur were kept burning in the halls.

Eventually, scientific research began to shed light on the etiology of YF and its link to mosquitoes. In 1881, Carlos Finlay, a Cuban physician, was the first to suggest that the disease was transmitted by a mosquito [68], but little was done to evaluate the claim until the establishment of a US Army Commission (Board) in Cuba in 1900 which set out to systematically investigate YF. The United States had won control of Cuba at the end of the Spanish–American War in 1898, and accordingly, many US military personnel were stationed there. Because of outbreaks of YF, the Surgeon General established a "board for the purpose of pursuing scientific investigations with reference to the acute infectious diseases on the island of Cuba, giving special attention to the etiology and prevention of yellow fever." The board set about to perform bacteriological studies on patients and victims of YF (they thought YF was caused by a bacterium) and to explore the theory of insect transmission. Attempts to prove a bacterial etiology for YF were negative. In fact, the causative agent of YF was shown to pass through bacteriological filters, indicating a viral etiology. This was the first demonstration of a human disease caused by a virus [68]. The researchers associated with the YF board—Walter Reed, Dr. James Carroll, Dr. Aristides Agramonte, and Jesse W. Lazear—designed several amazing experiments to unravel the YF mystery systematically. In one set of experiments, they built a "fomite" house in which human volunteers had to sleep on cots and in bed clothes soiled with a liberal quantity of black vomit, urine, and fecal matter from recent victims. In subsequent tests, the volunteers had to sleep on pillows covered with towels soaked with blood of YF victims. All volunteers remained well. The fomite theory of YF transmission was now history. Next, they built a "mosquito house," which had two partitions separated by a screen. Mosquitoes that had previously been fed on sick patients were released into one side of the house where a susceptible host—John Moran—reclined, clothed in only a nightshirt. He was bitten repeatedly. Other susceptible volunteers were asked to sleep in the other side of the house (free from mosquitoes). Five days later Moran came down with classic YF, but the others remained healthy. The board then made a far-reaching conclusion: regardless of its cause, YF must be transmitted by mosquitoes and therefore can be managed by mosquito control and patient isolation techniques (Fig. 3.32). Interestingly, many of the brave volunteers (the ones who got sick or stayed in the fomite house) who participated in these YF experiments were eventually awarded a Congressional Medal of Honor.

**Fig. 3.32** Once scientists discovered that yellow fever was mosquito-borne, efforts were specifically targeted toward mosquito control. (National Library of Medicine photo)

### 3.5.3  Jungle Vs Urban YF Cycles

Jungle (sylvatic) YF occurs in Africa and Central and South America maintained among monkeys by forest or scrub mosquitoes. Public health officials must diligently monitor jungle YF activity as it may be bridged into urban areas. In 1995, the largest jungle YF epidemic in history recorded 422 cases with 213 deaths [55]. In Africa, key links in the jungle cycle are Cercopithecidae monkeys (includes redtailed monkeys) and, more rarely, the lesser bush baby, infected by tree canopy mosquitoes, such as *Aedes africanus* [60]. Other mosquito species may be involved, especially in bridging the sylvatic cycle to an urban cycle. The virus can be introduced to urban areas in several ways. Villagers may acquire the virus while working in the forest and then return home ill, or infected red-tailed monkeys may venture into villages looking for food and be bitten by peridomestic mosquitoes. In the Americas, the sylvatic hosts are in the family Cebidae—especially howler and spider monkeys. Vector mosquitoes maintaining the sylvatic cycle include many species of *Haemagogus* mosquitoes. These mosquitoes breed in tree holes within dense forests but will bite humans if given the opportunity—situations like woodcutters clearing forests for agriculture. The urban cycle begins as these sickened foresters go back to their villages. The urban cycle (human-to-human) in both Africa and the Americas is maintained by domestic *Ae. aegypti*, a very common day-flying mosquito that breeds around homes in artificial containers, such as water pots, pet dishes, discarded tires, clogged rain gutters, soda cans, and so forth.

### 3.5.4  Treatment and Prevention

There is no specific treatment for YF, but prevention and control of epidemics can be achieved by use of the live 17D vaccine. This vaccine is one of the most successful live attenuated vaccines known to science. It is highly immunogenic, has a very low incidence of clinical reactions, and confers long-lasting (possibly lifetime) immunity. The package insert should be consulted before vaccine administration for advice about specific restrictions or exclusions (e.g., pregnant women). Transmission of YF may also be interrupted by mosquito avoidance, control, and, specifically, destruction of *Ae. aegypti* breeding sites.

Understanding the dynamics of urban YF gives one a unique perspective of the historical aspects of the disease. Discovering the urban mosquito vector—the exploitable weak link—was the key to stopping epidemics. At first, no one knew what spread the disease; it was just known that it moved in waves from one place to another. Some physicians logically assumed that fomites (articles of bedding or clothing contaminated with the agent) must be the cause. Others believed that YF was spread by miasmatic (poisoned) air. Fear and panic ruled during epidemics.

People were often held at bay at gunpoint, prevented from entering towns. No one knew at the time that the real problem was mosquito breeding! Trash, neglected water pots, rain barrels, and the like were everywhere in cities affected by epidemics. It seems odd that something as simple as a clean-up campaign, combined with mosquito avoidance (screens or nets), could so drastically reduce human suffering and death owing to this terrible disease.

**A Personal Connection to Yellow Fever**

Mary Caroline "Molly" Wilson (below) was born in 1861 in Memphis, TN. During an outbreak of yellow fever in 1867 (this was not the major epidemic of 1878), her parents died, and both she and her sister were placed in an orphanage. She was later adopted by the McShan family from Nettleton, MS (the sister was not adopted). Molly Wilson eventually married Isaac Franklin Smith in 1879 in Verona, MS, which was this book author's (Jerome Goddard's) great, great grandfather. Family legend says that when the McShans left with Molly from the orphanage, her sister was standing outside waving goodbye and crying. No one knows whatever became of that sister. Why they did not adopt the other girl remains unknown.

## 3.6   Lymphatic Filariasis

### 3.6.1   Introduction and Medical Significance

Filariae are long, threadlike nematodes (nonsegmented, cylindrical, tapered-at-both-ends worms) that inhabit the human lymphatic system and/or subcutaneous and deep connective tissues. Only species affecting the lymphatic system are discussed here. Other species that may occasionally infect human tissues are included in the next section. In 2000, an estimated 120 million people were infected with about 40 million disfigured and incapacitated [69]. There are basically two types of this mosquito-borne disease, relating to the species of roundworm involved—Bancroftian filariasis and Brugian (sometimes also called Malayan) filariasis. A third type, Timoran filariasis (named after the island of Timor), is rare and limited to several islands of southeastern Indonesia, and will not be addressed in this chapter. Filariasis occurs over much of the tropical world, whereas the Brugian form is mostly confined to Southeast Asia (China, India, and Malaysia) (Figs. 3.33 and 3.34). Clinical symptoms include fever, lymphangitis, lymphadenitis, occasional abscess formation, and chronic obstructive manifestations. The obstructive signs, elephantiasis (limbs) and hydroceles (genitalia), are thought by many to be the inevitable end result of filariasis; however, these are actually uncommon complications. As for the threat of human filariasis in the United States, millions of immigrants have come to the United States in the last few decades, many originating from countries with endemic filariasis. It is reasonable to expect some risk of infecting local mosquitoes. Also, immigrants may present to healthcare providers with signs and symptoms of filariasis [70].

**Fig. 3.33** Approximate geographic distribution of Bancroftian filariasis

**Fig. 3.34** Approximate geographic distribution of Brugian filariasis

### 3.6.2 Clinical and Laboratory Findings

Not evéryone exposed to filarial infection develops symptoms or signs other than perhaps microfilaremia—defined as larval worms, or microfilariae, in the circulating blood. Others develop inflammatory manifestations, such as an acute localized inflammation (the skin may be erythematous and hot), lymphadenitis, lymphangitis, and fever. There may be accompanying chills, sweats, headache, anorexia, lethargy, myalgias, and arthralgias. Abscesses may arise in the inguinal and axillary lymphatic structures, distal extremities, or breasts. Inflammation of the testicles, epididymis, and spermatic cord may also result. In fact, scrotal involvement may result in hydrocele (this is separate from scrotal elephantiasis). Obstructive filariasis may occur when adult worms in the inguinal (or other) lymph nodes cause obstruction of lymphatic drainage resulting in the legs or scrotum swelling to grotesque proportions. As this condition continues, it becomes stabilized and hardened by fibrosis—a condition called elephantiasis. Elephantiasis generally only develops in a small number of people who have

**Fig. 3.35** Microscopic view of *Brugia malayi* microfilaria. (CDC photo by Dr. Lee Moore)

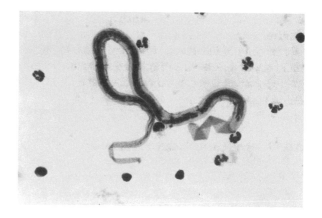

been exposed to filarial infections repeatedly over a period of years. Diagnosis of lymphatic filariasis is usually based on clinical exam, history of exposure in endemic areas, detection and identification of microfilariae in peripheral blood, and/or antibody tests, such as indirect fluorescent antibody (IFA) or enzyme-linked immunosorbent assay (ELISA). Polymerase chain reaction (PCR)-based assays to detect filarial DNA in blood are increasingly being used for diagnosis. Hypereosinophilia is a common laboratory finding. Microfilariae may be directly viewed microscopically (Fig. 3.35). Often, thin blood smears stained with Giemsa or Field's stain will reveal the microfilariae. In lysed thick blood films, microfilariae range in length from about 245 to nearly 300 μm and display a large number of distinct nuclei [71]. Each of the tiny worms is inside a thin, delicate sheath—the persisting egg membrane. Presence or absence of the sheath is important in the diagnosis (Fig. 3.35). Other species of nonpathogenic filariae may infect humans, which produce unsheathed microfilariae, but all pathogenic ones are sheathed. It must be noted that filarial infection may occur without detectable microfilaremia.

## 3.6.3  Ecology of Lymphatic Filariasis

### 3.6.3.1  Bancroftian Filariasis

Bancroftian filariasis, caused by *Wuchereria bancrofti*, is widely distributed through much of Central Africa, Madagascar, the Nile Delta, the Arabian seacoast, Turkey, India, Pakistan, Sri Lanka, Burma, Thailand, Southeast Asia, many Pacific islands, Malaysia, the Philippines, and the southern parts of China, Korea, and Japan. In the New World, it is found in the Caribbean and Central and South America, including Haiti, the Dominican Republic, Costa Rica, Honduras, Guatemala, Guyana, Suriname, French Guiana, and parts of Brazil. There was at one time a small endemic center of the disease near Charleston, SC, that has apparently disappeared. Bancroftian filariasis is an interesting disease in that there are no other known vertebrate hosts of the worms. It is transmitted solely by mosquitoes, and there is no multiplication of the parasite in the mosquito vector. The house mosquito, *Culex quinquefasciatus*, is a common urban vector; in rural areas, transmission is maintained mainly by *Anopheles* mosquitoes [72].

### 3.6.3.2   Brugian or Malayan Filariasis

Malayan filariasis does not occur in the New World, being mostly confined to Malaysia and areas from the Indian subcontinent through Asia to Japan. The disease has been virtually eradicated from Sri Lanka and Taiwan and nearly so from Mainland China. The life cycle of the causative agent, *Brugia malayi*, is similar to that of *W. bancrofti*, except that in most areas the principal mosquito vectors are in the genus *Mansonia*. However, *Anopheles* mosquitoes may also be involved. Some forms of Brugian filariasis may involve animal reservoirs, such as cats, monkeys, and pangolins.

### 3.6.3.3   Filarial Life Cycle in Hosts

Microfilariae are ingested in blood when mosquitoes feed on infected persons (Fig. 3.36). They penetrate the mosquito stomach wall, entering the body cavity (hemocoel), where they migrate to flight muscles for growth. After two molts, the third-stage infective larvae migrate through the head, eventually reaching the proboscis of the mosquito. By this time, the larvae are 1.5–2.0 mm long. During the mosquito's next blood meal, infective larvae escape onto human skin, where they enter through the mosquito bite puncture wound or local abrasions. In humans, the parasites pass to the lymphatic system where they undergo further molts eventually to become adult worms (several months later). Adult worms may live in humans—almost continuously producing thousands of microfilariae per day—for 10–18 yr [72].

### 3.6.3.4   Nocturnal Periodicity

In many areas of the world where filariasis occurs, the infection is seen in a "periodic form" wherein the microfilariae circulate in their animal or human hosts in higher numbers at night—supposedly being available in higher numbers when their specific mosquito vectors feed. Alternatively, the "subperiodic" form of filariasis describes the condition in which microfilaremias are roughly the same at all times. Much has been written about periodicity; it has classically been said that the proportion of the number of microfilariae found in blood smears during the day as opposed to night is 1:1000. However, periodicity may not be so pronounced and general as once thought, varying

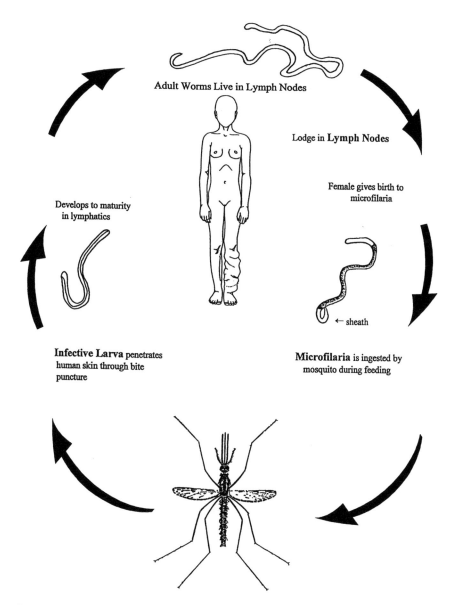

Adult Worms Live in Lymph Nodes

Lodge in **Lymph Nodes**

Develops to maturity
in lymphatics

Female gives birth to
microfilaria

← sheath

**Infective Larva** penetrates
human skin through bite
puncture

**Microfilaria** is ingested by
mosquito during feeding

**Fig. 3.36** Life cycle of *W. bancrofti*. (Provided with permission by Infections in Medicine 1998;15:607)

a great deal from place to place and species of filarial worm. Nonetheless, Bancroftian filariasis generally displays periodicity over much of its range (a subperiodic form does occur, however), whereas Malayan filariasis has several forms—nocturnal periodic, nocturnal subperiodic, and possibly a diurnal subperiodic form.

### 3.6.3.5  Treatment

Diethylcarbamazine (DEC), sometimes known as Hetrazan, has been used for years for the treatment of lymphatic filariasis. DEC is an extremely effective microfilaricide for *Wuchereria* and *Brugia* species, resulting in near-zero microfilaremia levels within hours, and one or more treatment courses may kill most of the adult worms. DEC is not widely available; contact the Centers for Disease Control Drug Service or the Division of Parasitic Diseases for availability and current treatment regimens. DEC is contraindicated in pregnant women and persons with renal disease. There may be systemic allergic reactions associated with rapid clearance of the microfilariae, but they usually can be managed with antipyretics, antihistamines, and analgesics. Over the last decade, the veterinary drug, ivermectin (Mectizan), has shown great promise in treatment of various filarial diseases, such as onchocerciasis. This drug may have an advantage over DEC in its efficacy when given as a single dose. However, ivermectin only kills the microfilariae, and its use in lymphatic filariasis is considered investigational at this time [73, 74].

## 3.7  Other Human-Infesting Filarial Worms

Numerous filarial worms are transmitted to humans and other mammals by mosquitoes and black flies. Examples include the causative agents of Bancroftian and Malayan filariasis (discussed above), loiasis, onchocerciasis, and dirofilariasis (dog heartworm). Other filarial worms may or may not cause symptomatic disease and are less well known (and thus have no common name), such as *Mansonella ozzardi*, *Mansonella streptocerca*, *Mansonella perstans*, *Dirofilaria tenuis*, *Dirofilaria ursi*, *Dirofilaria repens*, and others.

The dog heartworm, *Dirofilaria immitis*, occurs mainly in the tropics and subtropics but also extends into southern Europe and North America. This worm infects several canid species, sometimes cats, and, rarely, humans. Numerous mosquito species are capable of transmitting dog heartworm, especially those in the genera *Aedes*, *Anopheles*, *Aedes*, and *Culex*. Mosquitoes pick up the microfilariae with their blood meal when feeding on infected dogs. In endemic areas, a fairly high infection rate may occur in local mosquitoes.

Undoubtedly, thousands of people in the United States are bitten each year by mosquitoes infected with *D. immitis*. Fortunately, humans are accidental hosts, and the larvae usually die. However, they may occasionally be found as a subadult worm in the lung (seen as a coin lesion on X-ray exam) [75]. The incidence of dog heartworm in humans may well be decreasing in the United States because of widespread—and fairly consistent—treatment of domestic dogs for heartworm. The closely related *D. tenuis* is commonly found in the subcutaneous tissues of raccoons (again, mosquito-transmitted) and may accidentally infest humans as nodules in subcutaneous tissues.

# References

1. Harwood RF, James MT. Entomology in human and animal health. 7th ed. New York: Macmillan; 1979. p. 548.
2. Navy. Navy Medical Department guide to malaria prevention and control. Norfolk: U.S. Naval Environmental Health Center; 1984. p. 90.
3. Anonymous. Is Malaria elimination within reach? Lancet Infect Dis. 2017;17:461.
4. Baird JK. Telling the human story of Asia's invisible malaria burden. Lancet. 2017; 389:781–2.
5. Greenwood B. Progress with the PfSPZ vaccine for malaria. Lancet Infect Dis. 2017; 17:463–4.
6. WHO. World malaria report. World Health Organization, http://www.who.int/malaria/publications/world_malaria_report_2014/en/. 2014.
7. Cunnion S, Dickens T, Ehrhardt D, Need J, Wallace J. Navy medical department guide to malaria prevention and control. Norfolk: U.S. Navy Environ Hlth Cntr; 1984.
8. Gilles HM, Warrell DA. Bruce-Chwatt's essential Malariology. London: Arnold Publishers; 1993. p. 340.
9. Breman JG, Malaria SRW. In: Last JM, Wallace RB, editors. Public health and preventive medicine. 13th ed. Norwalk: Appleton and Lange; 1992. p. 1212–400.
10. Jensen T, Dritz DA, Fritz GN, Washino RK, Reeves WC. Lake Vera revisited: parity and survival rates of *Anopheles punctipennis* at the site of a malaria outbreak in the sierra Nevada foothills of California. Am J Trop Med Hyg. 1998;59(4):591.
11. Lluberas MF. Nothing but net only works in basketball. Winter: Wing Beats Magazine; 2007. p. 22–7.
12. Roberts DR, Manguin S, Mouchet J. DDT house spraying and re-emerging malaria. Lancet. 2000;356:330–2.
13. Anonymous. What's next for the malaria RTS,S vaccine candidate? Lancet. 2015;386:1708.
14. Clemens J, Moorthy V. Implementation of RTS,S/AS01 malaria vaccine–the need for further evidence. N Engl J Med. 2016;374:2596–7.
15. Breman JG, Alilio MS, Mills A. Conquering the intolerable burden of malaria: what's new, what's needed: a summary. Am J Trop Med Hyg. 2004;71(2 Suppl):1–15.
16. Heymann DL, editor. Control of communicable diseases manual. 20th ed. Washington, DC: American Public Health Association; 2015.
17. Goddard J, Currier M. Case histories of insect- or arachnid-caused illness. J Agromedicine. 1995;2:53–61.
18. Goddard J. Physician's guide to arthropods of medical importance. 6th ed. Boca Raton: Taylor and Francis Group (CRC Press); 2013. p. 515.
19. Cupp EW, Klingler K, Hassan HK, Viguers LM, Unnasch TR. Transmission of eastern equine encephalomyelitis virus in Central Alabama. Am J Trop Med Hyg. 2003;68(4):495–500.
20. Chamberlain RW. History of St. Louis encephalitis. In: Monath TP, editor. St Louis encephalitis. Washington, D.C.: American Public Health Association; 1980. p. 3–61.
21. Brinker KR, Monath TP. SLE: the acute disease. In: Monath TP, editor. St Louis encephalitis. Washington, D.C.: American Public Health Association; 1980. p. 503–34.
22. Tsai TF, St MCJ. Louis encephalitis. In: Monath TP, editor. The Arboviruses: epidemiology and ecology, vol. 4. Boca Raton: CRC Press; 1988. p. 113–43.
23. Shaman J, Day JF, Stieglitz M. Drought-induced amplification of Saint Louis encephalitis virus, Florida. Emer Infect Dis. 2002;8:575–80.
24. CDC. Outbreak of West Nile-like viral encephalitis in New York. CDC. MMWR. 1999;48:845–8.
25. Hayes EB, Komar N, Nasci RS, Montgomery SP, O'Leary DR, Campbell GL. Epidemiology and transmission dynamics of West Nile virus disease. Emerg Infect Dis. 2005;11:1167–73.
26. Mostashari F, Bunning ML, Kitsutani P. Epidemic West Nile encephalitis, New York, 1999: results of a household-based seroepidemiological survey. Lancet. 2001;358:261–4.

27. Asnis DW, Conetta R, Teixeira A. The West Nile virus outbreak of 1999 in New York: the flushing hospital experience. Clin Infect Dis. 2000;30:413–7.
28. Kilpatrick AM, Kramer LD, Campbell SR, Alleyne EO, Dobson AP, Daszak P. West Nile virus risk assessment and the bridge vector paradigm. Emerg Infect Dis. 2005;11(3):425–9.
29. Molaei G, Andreadis TG, Armstrong PM, Anderson JF, Vossbrinck CR. Host feeding patterns of *Culex* mosquitoes and West Nile virus transmission, northeastern United States. Emerg Infect Dis. 2006;12:468–74.
30. CDC. Western equine and other encephalitis case numbers. In: Arbovirus diseases branch, division of vector-borne infectious diseases. Ft. Collins; 1998.
31. Turell MJ. Vector competence of three Venezuelan mosquitoes for an epizootic IC strain of Venezuelan equine encephalitis virus. J Med Entomol. 1999;36:407–9.
32. USDA. Venezuelan equine encephalomyelitis, a national emergency. USDA, Animal Plant Health Inspection Service, Washington, DC; 1972. APHIS-81-1
33. Tesh RB, Solomon T. Japanese encephalitis, West Nile, and other flavivirus infections. In: Guerrant RL, Walker DH, Weller PF, editors. Tropical infectious diseases. 3rd ed. New York: Saunders Elsevier Publishing; 2011.
34. CDC. Japanese encephalitis surveillance and immunization–Asia and the western Pacific. CDC. MMWR. 2013;62:658–62.
35. Burke DS, Leake CJ. Japanese encephalitis. In: Monath TP, editor. The Arboviruses: Epidemiology and Ecology. Vol. 3. Boca Raton: CRC Press, Inc.; 1988. p. 63–92.
36. Mannix FL, Wesson DW, Potential for introduction and establishment of Japanese encephalitis virus in North America. Presentation at the 56th annual meeting of the American Society of Tropical Medicine and Hygiene, November 4–8. 2007.
37. Enserink M. Tropical disease follows mosquitoes to Europe. Science (News Focus). 2007;317:1485.
38. Weaver SC, Smith DW. Alphavirus infections. In: Guerrant RL, Walker DH, Weller PF, editors. Tropical infectious diseases. 3rd ed. London: Saunders (Elsevier); 2011. p. 519–24.
39. Enserink M. Chikungunya: no longer a third world disease. Science (News Focus). 2007;318:1860–1.
40. Leparc-Goffart I, Nougairede A, Cassadou S, Prat C, de Lamballerie X. Chikungunya in the Americas. Lancet. 2014;383:514.
41. Summary CDC. Of notifiable infectious diseases and conditions -- United States, 2015. CDC. MMWR. 2017;64(53):1–144.
42. Plourde AR, Bloch EM. A literature review of Zika virus. Emerg Infect Dis. 2016;22(7): 1185–92.
43. Petersen LR, Jamieson DJ, Honein MA. Zika virus. N Engl J Med. 2016;375(3):294–5.
44. Gatherer D, Kohl A. Zika virus: a previously slow pandemic spreads rapidly through the Americas. J Gen Virol. 2016;97:269–73.
45. CDC. Cumulative Zika virus disease case counts in the United States, 2015-2017. CDC website, https://www.cdc.gov/zika/reporting/case-counts.html. 2017.
46. Lambrechts L, Scott TW, Gubler DJ. Consequences of the expanding global distribution of *Aedes albopictus* for dengue virus transmission. PLoS Negl Trop Dis. 2010;4(5):e646.
47. Rai KS. *Aedes albopictus* in the Americas. Annu Rev Entomol. 1991;36:459–84.
48. Gubler DJ. Epidemic dengue and dengue hemorrhagic fever: a global public health problem in the 21st century. In: Scheld WM, Armstrong D, Hughes JM, editors. Emerging infections. Vol. 1. Washington, D.C.: ASM Press; 1998. p. 1–14.
49. Dengue SAM. An underappreciated threat. Inf Med. 2005;22:304–6.
50. Anonymous. Dengue more prevalent than thought. Science (News Focus). 2013;340:127.
51. CDC. Locally acquired dengue -- Key West, Florida, 2009-2010. CDC. MMWR. 2010;59:577–81.
52. Hahn MB, Eisen L, McAllister J, Savage HM, Mutebi JP, Eisen RJ. Updated reported distribution of *Aedes (Stegomyia) aegypti* and *Aedes (Stegomyia) albopictus* (Diptera: Culicidae) in the United States, 1995-2016. J Med Entomol. 2017;54(5):1420–4.
53. CDC. Dengue hemorrhagic fever -- U.S.-Mexico border, 2005. CDC. MMWR. 2007;56:785–9.
54. CDC. Surveillance summary. CDC. MMWR. 1994;43/SS-2:8. 1994

55. Halstead SB. Emergence mechanisms in yellow fever and dengue. In: Scheld WM, Craig WA, Hughes JM, editors. Emerging infections, part II, vol. 2. Washington, DC: ASM Press; 1998. p. 65–79.

56. Normile D. Hunt for dengue vaccine heats up as the disease burden grows. Science (News Focus). 2007;317:1494–5.

57. Normile D. Dengue vaccine trial poses public health quandry. Science (News Focus). 2014;345:367.

58. Wilder-Smith A, Gubler DJ. Dengue vaccines at a crossroad. Science (News Focus). 2015;350:626–7.

59. Normile D. Safety concerns derail dengue vaccination program. Science (News Focus). 2017;358:1514–5.

60. Service MW. Mosquitoes. In: Lane RP, Crosskey RW, editors. Medical insects and arachnids. London: Chapman and Hall; 1996. p. 120–240.

61. Sanders EJ, Borus P, Ademba G, Kuria G, Tukei PM, LeDuc JW. Sentinel surveillance for yellow fever in Kenya. Emerg Infect Dis. 1996;2:236–8.

62. Garske T, Van Kerkhove MD, Yactayo S, Ronveaux O, Lewis RF, Staples JE, et al. Yellow fever in Africa: estimating the burden of disease and impact of mass vaccination from outbreak and serological data. PLoS Med. 2014;11(5):e1001638.

63. Barrett AD. Yellow fever in Angola and beyond–the problem of vaccine supply and demand. N Engl J Med. 2016;375:301–3.

64. MacFarland JM, Baddour LM, Nelson JE. Imported yellow fever in a United States citizen. Clin Infect Dis. 1997;25:1143–7.

65. Cope SE. Yellow fever–the scourge revealed. Florida Mosquito Control Association, Wing Beats. 1996;7:14–26.

66. Crosby MC. The American plague. New York: Berkley Books; 2006. p. 308.

67. White M. Yellow fever. Memphis: The Commercial Appeal Newspaper; 1978.

68. Bres PLJ. A century of progress in combating yellow fever. Bull WHO. 1986;64:775–86.

69. WHO. Fact sheet: lymphatic filariasis. Geneva: World Health Organization; 2017. p. 4.

70. Cunningham NM. Lymphatic filariasis in immigrants from developing countries. Am Fam Phys. 1997;55:119–1204.

71. Markell E, Voge M, John D. Medical parasitology. 7th ed. Philadelphia: W.B. Saunders; 1992.

72. Brygoo ER. Epidemiology of filariasis. Noumea: Proceedings of a conference on filariasis in the South Pacific, South Pacific Commission; 1953.

73. Cao WC, Van der Ploeg CP, Van der Sluijs IJ, Habbema JD. Ivermectin for the chemotherapy of bancroftian filariasis: a meta-analysis of the effect of single treatment. Tropical Med Int Health. 1997;2:393–403.

74. de Silva N, Guyatt H, Anthelmintics BD. A comparative review of their clinical pharmacology. Drugs. 1997;53:769–88.

75. Thomas JG, Sundman D, Greene JN, Coppola D, Lu L, Robinson LA, et al. A lung nodule: malignancy or the dog heartworm? Inf Med. 1998;15:105–6.

# Chapter 4
# Tick-Borne Diseases

## 4.1 Basic Tick Biology

There are three families of ticks recognized in the world today:

1. Ixodidae (hard ticks)
2. Argasidae (soft ticks)
3. Nuttalliellidae (a small, curious, little-known group with some characteristics of both hard and soft ticks)

The terms hard and soft refer to the presence of a dorsal scutum or "plate" in the Ixodidae, which is absent in the Argasidae. Hard ticks display sexual dimorphism, whereby males and females look obviously different (Fig. 4.1), and the blood-fed females are capable of enormous expansion. Their mouthparts are anteriorly attached and visible from dorsal view. If eyes are present, they are located dorsally on the sides of the scutum.

Soft ticks are leathery and nonscutate, without sexual dimorphism (Fig. 4.2). Their mouthparts are subterminally attached in adult and nymphal stages and not visible from dorsal view. Eyes, if present, are located laterally in folds above the legs.

There are major differences in the biology of hard and soft ticks. Some hard ticks have a one host life cycle, wherein engorged larvae and nymphs remain on the host after feeding; after they molt, subsequent stages reattach and feed. Adults mate on the host, and only engorged females drop off to lay eggs on the ground. Although some hard ticks complete their development on only one or two hosts, most commonly encountered ixodids have a three-host life cycle (Fig. 4.3). In this case, adults mate on a host (except for some *Ixodes* spp.), and the fully fed female drops from the host animal to the ground and lays from 2000 to 18,000 eggs after which she dies. Eggs hatch in about 30 d into a six-legged seed tick (larval) stage, which feeds predominantly on small animals. The fully fed seed ticks drop to the ground and transform into eight-legged nymphs. These nymphs seek an animal host, feed, and drop to the ground. They then molt into adult ticks, thus completing the life cycle.

© Springer International Publishing AG, part of Springer Nature 2018
J. Goddard, *Infectious Diseases and Arthropods*, Infectious Disease,
https://doi.org/10.1007/978-3-319-75874-9_4

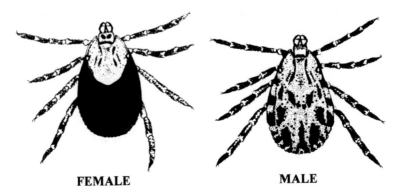

**FEMALE**                    **MALE**

**Fig. 4.1** Female and male hard ticks (Family Ixodidae). (From US Pub. Hlth Serv. NIH Bull. No. 171)

**Fig. 4.2** Soft tick (Family Argasidae) *Otobius megnini* (US Air Force photo)

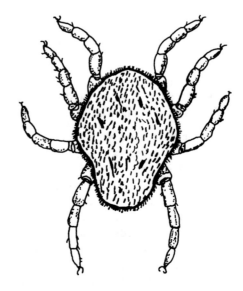

Many hard tick species "quest" for hosts, by climbing blades of grass or weeds and remaining attached, forelegs outstretched, awaiting a passing host. They may travel up a blade of grass (to quest) and back down to the leaf litter where humidity is high (to rehydrate) several times a day. Also, hard ticks will travel a short distance toward a $CO_2$ source. Adult ticks are more adept at traveling through vegetation than the minute larvae.

Ticks feed exclusively on blood and begin the process by cutting a small hole into the host epidermis with their chelicerae and inserting the hypostome into the cut, thereby attaching to the host. Blood flow is maintained with the aid of anticoagulants and vasodilators secreted by the salivary glands. Some hard ticks secure their attachment to the host by forming a cement cone around the mouthparts and surrounding skin. Two phases are recognized in the feeding of nymphal and female

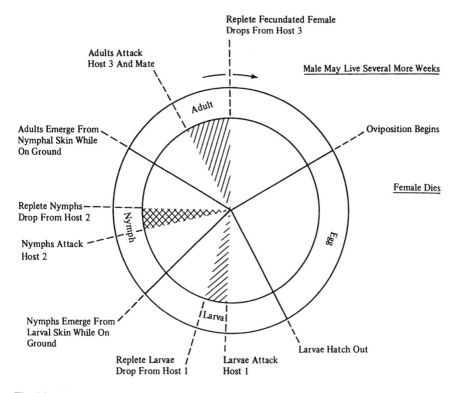

**Fig. 4.3** Life cycle of a three-host tick. (From USDA, ARS, Agri. Hndbk. No. 485)

hard ticks: a growth feeding stage characterized by slow continuous blood uptake and a rapid engorgement phase occurring during the last 24 h or so of attachment.

The biology of soft ticks differs from hard ticks in several ways. Adult female soft ticks feed and lay eggs several times during their lifetime. Soft tick species may also undergo more than one nymphal molt before reaching the adult stage. With the exception of larval stages of some species, soft ticks do not firmly attach to their hosts for several days like the Ixodidae—they are adapted to feeding rapidly and leaving the host promptly.

Hard ticks and soft ticks occur in different habitats. In general, hard ticks occur in brushy, wooded, or weedy areas containing numerous deer, cattle, dogs, small mammals, or other hosts. Soft ticks are generally found in animal burrows or dens, bat caves, dilapidated or poor-quality human dwellings (huts, cabins, and so forth), or animal rearing shelters. Many soft tick species thrive in hot and dry conditions, whereas ixodids are more sensitive to desiccation and, therefore, usually found in areas providing protection from high temperatures, low humidities, and constant breezes.

Most hard ticks, being sensitive to desiccation, must practice water conservation and uptake. Their epicuticle contains a wax layer, which prevents water movement through the cuticle. Water can be lost through the spiracles; therefore, resting ticks keep their spiracles closed most of the time, opening them only one or two times an hour. Tick movement and its resultant rise in $CO_2$ production cause the spiracles to open about 15 times an hour with a corresponding water loss.

Development, activity, and survival of hard ticks are influenced greatly by temperature and humidity within the tick microhabitat. Because of their temperature and high humidity requirements, as well as host availability, hard ticks tend to congregate in areas providing those factors. Ecotonal areas (interface areas between major habitat types) are excellent habitats for hard ticks. Open meadows/prairies, along with climax forest areas, generally support the fewest ticks (there are exceptions—the Gulf Coast tick seems to prosper in open fields). Ecotone areas and small openings in the woods are usually heavily infested. Deer and small mammals thrive in ecotonal areas, thus providing blood meals for ticks. In fact, deer are often heavily infested with hard ticks in the spring and summer months. The optimal habitat of white-tailed deer has been reported to be the forest ecotone, since the area supplies a wide variety of browse and frequently offers the greatest protection from their natural enemies. Many favorite deer foods are also found in the low trees of an ecotone, including greenbrier, sassafras, grape, oaks, and winged sumac.

Ticks are not evenly distributed in the wild; instead, they are localized in areas providing their necessary temperature, humidity, and host requirements. These biologic characteristics of ticks, when known, may enable us to avoid the parasites.

**Family Clusters of Rocky Mountain Spotted Fever**

Rocky Mountain spotted fever (RMSF) may sometimes occur in a cluster, confounding diagnosis because physicians may think the illness is viral or bacterial and transmitted from person to person. At least two such clusters have occurred in Mississippi. In one case, during April, a 5-yr-old child became febrile (101 °F oral) and irritable, with vomiting and diarrhea, and a day later developed a generalized macular rash. After two more days, she showed no improvement in her clinical status and was seen by a pediatrician. By this time, she had developed nuchal rigidity and became disoriented.

The pediatrician noted that the rash had become petechial and suspected RMSF. She was immediately admitted to the hospital where i.v. chloramphenicol therapy was begun. Over the next three hours, her clinical and neurologic status deteriorated precipitously, and arrangements were made to transfer her to a regional medical center. However, 10 min after leaving the hospital by ambulance, the child had a cardiopulmonary arrest and was rushed back to the hospital emergency department (ED) while resuscitation was attempted. The effort was unsuccessful and death was pronounced. Three days after her death, the patient's 16-yr-old sister presented to the ED with fever, headache, and a generalized macular rash involving her palms and soles. She was admitted to the hospital and empiric therapy with tetracycline was started. Her symptoms completely resolved after 3 days.

(continued)

In another family cluster of RMSF, a husband and wife developed symptoms consistent with RMSF. The diagnosis was unsuspected in the man's case and came too late for effective treatment in his wife's case. Both patients died. Fortunately, the correct diagnosis was made in the woman's case in time to intervene with effective tetracycline therapy and cure her son, who acquired his RMSF after coming to visit for the funeral of his parents. Investigation disclosed ticks in the couple's house, including one in the man and woman's bed; investigators concluded that the ticks came from a pet dog who often slept in their bedroom.

From Conwill DE, Oakes T, Brackin BT. Mississippi Morbidity Report. 5;1987:1–2.

## 4.2 Rocky Mountain Spotted Fever (RMSF)

### 4.2.1 Introduction

Rickettsiae are small, obligate intracellular bacteria. For classification purposes, they are grouped into several broad categories, such as the spotted fever group, the typhus group, and (some researchers say) the scrub typhus group. In addition to the genus *Rickettsia*, several closely related members of the family Anaplasmataceae (*Ehrlichia* and *Anaplasma*) may also cause human diseases (*see* Sects. 4.5 and 4.6 in this chapter). There are many rickettsial species in the spotted fever group (SFG); it contains at least 20 disease agents and another 20 or so with low or no pathogenicity to humans [1]. Table 4.1 presents some distributional and epidemiologic information on 14 of the human disease-causing SFG rickettsiae. Most are clinically similar. For example, American boutonneuse fever cases are often diagnosed as RMSF in the United States, and for that reason, the CDC changed their reporting category to "spotted fever rickettsiosis.

As mentioned, not all SFG rickettsiae are pathogenic. Numerous SFG species can be isolated from field-collected ticks [2, 3]. This often leads to confusion, since they look alike microscopically and will react with fluorescent antibody stains (Fig. 4.4). Historically, research studies indicated a percentage of SFG-positive ticks found in an area, but without further differentiation regarding species by PCR or other means, this information is almost useless. Just because a SFG rickettsia occurs in ticks in a park, for example, does not necessarily mean there is a threat from RMSF or any other SFG pathogen. Much is unknown about these nonpathogenic rickettsial organisms, and some researchers contend that all of them could potentially cause disease in humans. Furthermore, the role they play in the ecology of human pathogens is complex, such as in the case of RMSF wherein these nonpathogenic species may interfere with the cycle of *Rickettsia rickettsii* by infecting ticks and thus crossprotecting them from the true *R. rickettsii* [4] (*see* Chap. 2). The remainder of this section will be limited to RMSF.

**Table 4.1** Epidemiologic information on 14 spotted fever group rickettsiae

| Rickettsia | Disease | Tick/mite vectors | Distribution |
|---|---|---|---|
| R. rickettsii | RMSF | Primarily ticks *Dermacentor variabilis, D. andersoni*, and *Rhipicephalus sanguineus* | Western Hemisphere |
| R. conorii | Boutonneuse fever or Mediterranean spotted fever | Primarily ticks in the genera *Rhipicephalus, Hyalomma*, and *Haemaphysalis* | Africa, Mediterranean region, Middle East |
| R. parkeri | American boutonneuse fever | Ticks, *Amblyomma maculatum, A. triste* | United States (*A. maculatum*) South America (*A. triste*) |
| R. africae | African tick bite fever | Ticks *Amblyomma hebraeum* and *A. variegatum* | Sub-Saharan Africa |
| R. sibirica | North Asian tick typhus (Siberian tick typhus) | Primarily ticks in genera *Dermacentor* and *Hyalomma* | Siberia, Central Asia, Mongolia |
| R. australis | Queensland tick typhus | Tick *Ixodes holocyclus* | Eastern Australia |
| R. slovaca | Tick-borne lymphadenopathy (TIBOLA) | Ticks, primarily *Dermacentor marginatus* | Europe |
| R. akari | Rickettsialpox | Mite, *Liponyssoides sanguineus* | United States and Mexico, possibly worldwide |
| R. japonica | Japanese spotted fever | Ticks, probably *Haemaphysalis flava, H. longicornis, Ixodes ovatus* | Japan, Korea |
| R. honei | Flinders Island spotted fever | Ticks in genera *Ixodes, Rhipicephalus*, and *Amblyomma* | Australia, Southeast Asia |
| R. phillipi (364D agent) | Pacific Coast tick fever | Tick *Dermacentor occidentalis* | Western United States (California) |
| R. amblyommatis | Not named | Tick *Amblyomma americanum* | Eastern United States |
| R. aeschlimannii | Not named | Tick, *Hyalomma marginatum* | Africa |
| R. helvetica | Not named | Tick *Ixodes ricinus* | Europe |

## 4.2.2   History

Major Marshall Wood, an Army physician, first identified a RMSF patient in Idaho in 1896, and case descriptions from Idaho and Montana were soon published. The disease gets its name because outbreaks of RMSF were first recognized in Montana, Idaho, Nevada, and Wyoming [5]. In 1904–1906, Howard Ricketts, for whom the pathogen is named, conducted groundbreaking research which identified the causative agent, its vector, and mode of transmission [6]. In a hyperendemic area in the

**Fig. 4.4** Fluorescent antibody stain of Gulf Coast tick salivary gland showing spotted fever group rickettsiae, but not the agent of RMSF R. rickettsia. (Photo courtesy Dr. Kristine Edwards, Mississippi State University)

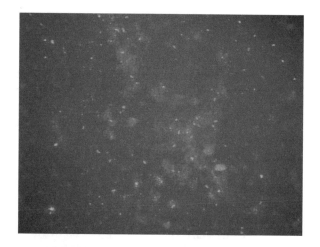

**Fig. 4.5** Tick researchers at the Rocky Mountain Laboratories developed a crude but effective vaccine against RMSF (National Library of Medicine photo)

Bitterroot Valley of Montana, the United States Public Health Service and the Montana Board of Health established a research station that eventually became the Rocky Mountain Laboratories (RML). During the 1920s RML scientists produced a crude, but effective, RMSF vaccine from ground-up ticks [7] (Fig. 4.5), which, about a decade later, was replaced by a vaccine prepared from rickettsiae grown in egg yolk sacs. The RML was crucial in early rickettsiology research, and has subsequently produced fundamental research in many areas of arthropod-transmitted diseases, perhaps most notably the discovery of the agent of Lyme disease by Dr. Willy Burgdorfer in 1982 [8].

### 4.2.3   Clinical and Laboratory Aspects of RMSF

RMSF is the most frequently reported rickettsial disease in the United States with about 4000 cases reported each year [9]. RMSF incidence is apparently increasing, especially among Native Americans. Probably even more cases occur but go unreported. Why? If an unusual febrile illness is treated successfully with doxycycline, there may be little interest in follow-up and reporting. At the time of initial presentation, there is often the classic triad of RMSF—fever, rash, and history of tick bite. Other characteristics are malaise, severe headache, chills, and myalgias. Sometimes gastrointestinal symptoms, such as abdominal pain and diarrhea, are reported. I have seen the proper diagnosis missed because of GI involvement. The rash, appearing on about the fifth day, usually begins on the extremities and then spreads to the rest of the body. However, there have been confirmed cases without rash. Mental confusion, coma, and death may occur in severe cases. Untreated, the mortality rate is about 20%; even with treatment, the rate is 4% [10]. There have been mild to severe neurological sequelae following RMSF infection such as encephalopathy, seizures, paresis, and peripheral motor neuropathies [11].

Laboratory findings include a normal or depressed leukocyte count, thrombocytopenia, elevated serum hepatic aminotransferase levels, and hyponatremia, although these abnormalities are not specific for RMSF [12, 13]. Specific tests to diagnose RMSF are not widely available (usually only through CDC or some universities). Indirect fluorescent antibody (IFA) tests on acute and convalescent sera are fairly accurate and can be used later to confirm the diagnosis. Older technologies such as Weil–Felix reactions with Proteus OX-19 and OX-2 were used in the past but now are considered unreliable. PCR on blood or tissue samples can also be used for diagnosis. RMSF organisms may be visualized in postmortem samples by immunohistochemistry (IHC).

### 4.2.4   Ecology of RMSF

RMSF is usually transmitted by the bite of an infected tick. Not all tick species are effective vectors of the rickettsia, and even in the vector species, not all ticks are infected. Therefore, tick infection with *R. rickettsii* is like a needle in a haystack. Generally, only 1–5% of vector ticks in an area are infected. Several tick vectors may transmit RMSF organisms, but the primary ones are the American dog tick *Dermacentor variabilis* in the eastern United States, and *Dermacentor andersoni* in the West (Figs. 4.6, 4.7, 4.8, and 4.9). The brown dog tick, *Rhipicephalus sanguineus*, is also a confirmed vector [14]. Adults of both *Dermacentor* species feed on a variety of medium to large mammals and humans [15, 16]. Ticks are often brought into close contact with people via pet dogs or cats (dog ticks may also feed on cats). In one case I investigated, the mother of the 3-yr-old patient said, "He always carried that puppy around… holding it up next to his face." Another mode of RMSF transmission may be manual deticking of dogs and subsequent autoinfection via mucosal membranes or eyes. One man contracted RMSF in Mississippi by

**Fig. 4.6** Adult female *D. andersoni.* (Photo courtesy Dr. Balke Layton, Mississippi State University)

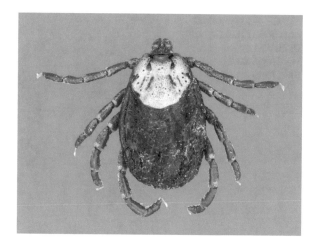

**Fig. 4.7** Approximate geographic distribution of *D. andersoni*

**Fig. 4.8** Adult female *D.*
*variabilis.* (Photo courtesy
Dr. Balke Layton,
Mississippi State
University)

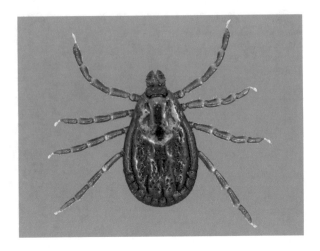

**Fig. 4.9** Approximate geographic distribution of *D. variabilis*

biting ticks, removed from his dog, between his teeth. That may seem odd, but I have since encountered other persons who claimed to kill ticks by "biting them."

### 4.2.5  Prevention and Treatment of RMSF

The only sure way of preventing tick-borne diseases is to prevent tick bites. Personal protection techniques for tick bites include avoiding tick-infested woods if possible, tucking pants legs into boots or socks, and using repellents on pant legs and socks (see Appendix 3). Products containing the active ingredients DEET or picaridin work fairly well in repelling ticks, but permethrin products are more effective. In addition, inspection of the body and removal of attached ticks are more important than many people realize (Fig. 4.10). In most tick-borne diseases, there is a feeding period required before transmission of the disease agent occurs. RMSF organisms generally take 1–3 h for transmission to occur. Doxycycline is the drug of choice for treatment of suspected or confirmed cases of RMSF in adults and children. Children under eight and pregnant women are sometimes given chloramphenicol, although there is no good reason why children shouldn't be given doxycycline for RMSF [17, 18]. Treatment should be initiated on clinical and epidemiologic grounds and *never* while waiting for confirmation of diagnosis [10, 19].

## 4.3  American Boutonneuse Fever (ABF)

### 4.3.1  Introduction and Background

Several years ago, investigators at the CDC discovered a new tick-borne disease or, more accurately, a "disease within a disease," because the new clinical entity was apparently hidden within cases diagnosed as Rocky Mountain spotted fever (RMSF). There are numerous spotted fever group (SFG) rickettsial species associated with ticks in the United States (*see* Table 4.1), but until about 20 years ago, only one was conclusively proved to be a human pathogen.

Over the course of several decades, rickettsiologists speculated about the role of so-called nonpathogenic rickettsiae in human disease, especially *Rickettsia parkeri* [20–22], which had been shown to cause mild clinical signs in guinea pigs and even eschar-like necrosis at sites of tick attachment [23]. Evidence that this species could cause illness in humans was provided when Paddock and associates [24] isolated *R. parkeri* from a patient with suspected rickettsialpox who was evaluated at the Portsmouth Naval Medical Center, Portsmouth, VA. This was the first report of human rickettsiosis caused by *R. parkeri*; many others have followed [25–28].

Because of the clinical similarities between disease caused by *R. parkeri* and an illness in Europe and Africa termed "boutonneuse fever" (caused by a closely related organism, *Rickettsia conorii*), a good descriptive moniker for this newly recognized rickettsiosis is "American boutonneuse fever" [29].

**Fig. 4.10** Recommended
method for tick removal:
grasp tick with forceps
near "head" region and
pull straight off. Do not
turn or twist. Disinfect bite
site. (From US Army
Center for Health
Promotion and Preventive
Medicine, "Focus on Lyme
Disease" Issue 9, Fall,
1996)

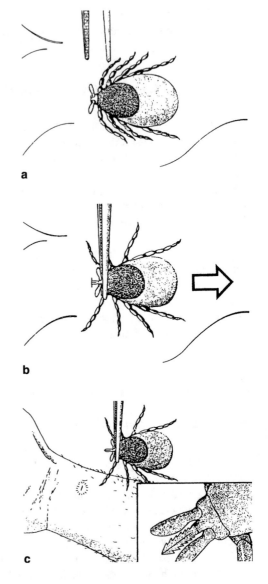

### 4.3.1.1   Clinical and Laboratory Description of a Case of ABF

A 40-yr-old man from Virginia complained of fever, mild headache, malaise, diffuse myalgias and arthralgias, and multiple eschars (spots of skin necrosis) on his
lower extremities [24]. Four days earlier, he had noted three papules on his lower
leg that developed into pustules, then ulcerated. Two days after he noticed the papules, systemic symptoms and a fever (temperature up to 39.2 °C [102.5 °F]) developed. The patient had recently walked his dogs in a grassy field but did not recall
any tick or mite bites.

Treatment was begun with amoxicillin and clavulanic acid for secondarily infected arthropod bites, but the patient's symptoms did not improve. Within 12 h, a mildly pruritic, erythematous, maculopapular rash developed on his trunk and soon spread to involve the palms and soles. Antibiotic therapy was changed to cephalexin, but symptoms persisted.

An infectious disease consultation was obtained. A faint, diffuse, salmon-colored rash was observed, predominantly on the abdomen (with some lesions on the extremities, hands, and face), along with a few scattered pustules. Three eschars, 1.5 cm in diameter, were identified on the pretibial aspect of the lower legs. Rickettsialpox was diagnosed, and treatment with doxycycline was initiated. The patient's fever, arthralgias, and myalgias resolved within 3 d, and the rash resolved within 1 wk.

Serologic evaluation by the CDC revealed antibodies reactive with *Rickettsia akari* and *Rickettsia rickettsii*, and SFG rickettsiae were visualized by immunohistochemical staining of a skin biopsy specimen. Subsequently, an SFG rickettsia was isolated in Vero cell culture from a second skin biopsy specimen. Molecular analyses of the isolate from the biopsy specimen examining multiple rickettsial genes confirmed the identity of the spotted fever rickettsia as *R. parkeri*.

### 4.3.2 Ecology of ABF

Many gaps remain in our knowledge about the natural history and ecology of *R. parkeri*. In the United States, the agent has been identified thus far from only three species of ticks—the Gulf Coast tick, the lone star tick, and an obscure rabbit tick [2, 30]—so any of them could theoretically be a vector. The Gulf Coast tick and lone star tick are both found in Virginia, where the patient acquired his infection, although the Gulf Coast tick would probably be less common in that area than the lone star tick.

The Gulf Coast tick (Fig. 4.11) is found along the Atlantic and Gulf Coast areas (generally 100–200 miles inland) and south into Mexico and portions of Central and South America. Currently this species is expanding inland and northward in the United States. The lone star tick occurs over much of the eastern and south central United States. Both species are large, fast-moving ticks that aggressively bite humans [31]. In addition, they both have immature stages (sometimes called "seed ticks") during which they feed on a variety of animals and ground-frequenting birds. Interestingly, peak seed tick activity is in August, the month in which the index patient became ill.

As for animal reservoirs of *R. parkeri*, any animal or bird on which the ticks frequently feed could theoretically serve as a reservoir. Candidates include cotton rats, meadowlarks, and bobwhite quail [32]. Alternatively, the ticks themselves may be the reservoir, with transovarial and transstadial transmission of the agent occurring indefinitely.

**Fig. 4.11** Adult female
*Amblyomma maculatum*,
primary vector of ABF.
(Photo courtesy Dr. Blake
Layton, Mississippi State
University)

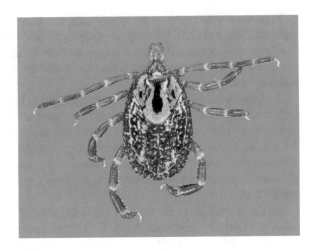

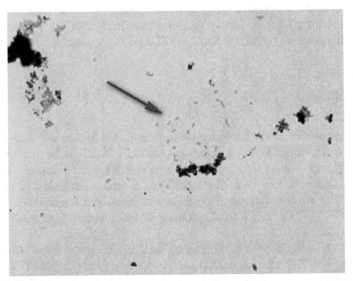

**Fig. 4.12** Rickettsia parkeri in spot of tick blood (hemolymph). (Photo courtesy Dr. Kristine
Edwards, Mississippi State University)

The coexistence of multiple tick-borne SFG rickettsioses sharing common geographic distributions has been reported previously in southern Europe and Africa
[33, 34]. The identities of individual and unique rickettsial agents may be obscured
when standard serologic assays or group-specific immunohistochemical staining
methods are used to confirm the diagnosis of an SFG rickettsiosis. As mentioned,
several SFG rickettsiae infect ticks (Fig. 4.12), and extensive antigenic cross-
reactivity exists among SFG rickettsiae. Therefore, most currently available tests
are only group-specific and cannot be used to identify a particular species [35]. The
cases reported by Paddock and associates [24, 26, 27] demonstrate that establishing

definitive causative associations for SFG rickettsiae is greatly facilitated by clinical foresight and collection of appropriate diagnostic specimens during evaluation of patients with febrile, eschar-associated illnesses.

### 4.3.3   Prevention and Treatment of ABF

Prevention and treatment of ABF are the same as that for Rocky Mountain spotted fever (*see* above section).

## 4.4   Other Spotted Fever Group Rickettsioses

### 4.4.1   Boutonneuse Fever

Boutonneuse fever (BF), or Mediterranean spotted fever, caused by *Rickettsia conorii*, is widely distributed in Africa, areas surrounding the Mediterranean, southern Europe, and India. The name is derived from the black, button-like lesion (eschar) at the site of tick bite (Fig. 4.13). BF resembles a mild form of RMSF, characterized by

**Fig. 4.13** Typical eschar at site of tick bite (Armed Forces Institute of Pathology negative no. D.4451)

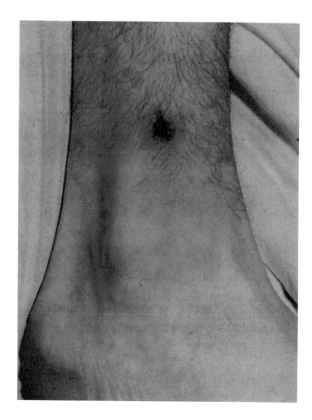

**Fig. 4.14** *A. hebraeum*, vector of boutonneuse fever and African tick bite fever (USAF photo B. Burnes)

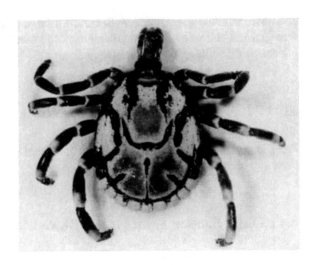

mild to moderately severe fever and a rash usually involving the palms and soles. Several tick species serve as vectors of the agent to humans, but especially *Rhipicephalus sanguineus*, *R. appendiculatus*, and *Amblyomma hebraeum* (Fig. 4.14).

## 4.4.2   African Tick Bite Fever

African tick bite fever (ATBF), caused by *Rickettsia africae*, is clinically similar to BF with the exception that there is usually an absence of rash (or just a transient rash) in ATBF patients. The disease is mild and is characterized by headache, fever, eschar at the tick bite site, and regional lymphadenopathy. ATBF primarily occurs in sub-Saharan Africa (particularly South Africa), where it is transmitted by various *Amblyomma* ticks, especially *A. hebraeum* (Fig. 4.14). Raoult and Olson [1] believe that ATBF is the most prevalent of the rickettsioses in the world.

## 4.4.3   Siberian Tick Typhus

Siberian tick typhus (STT), or North Asian tick typhus, caused by *Rickettsia sibirica*, is very similar clinically to RMSF with fever, headache, and rash. The disease can be mild to severe but is seldom fatal. STT was first recognized in the Siberian forests and steppes in the 1930s but now is known to occur in many areas of Asiatic Russia and on islands in the Sea of Japan. Various hard ticks are vectors of the agent, but especially *Dermacentor marginatus*, *D. silvarum*, *D. nuttalli*, and *Haemaphysalis concinna*.

### 4.4.4 Queensland Tick Typhus

Queensland tick typhus (QTT), caused by *Rickettsia australis*, occurs in eastern Australia. It is primarily restricted to dense forests interspersed with grassy savanna or secondary scrub. Most patients have fever, headache, and rash that may be vesicular and petechial—even pustular. Commonly, there is an eschar at the site of tick bite. The agent of QTT is transmitted to humans by *Ixodes holocyclus* ticks.

## 4.5 Ehrlichiosis

### 4.5.1 Introduction

*Ehrlichia* are microorganisms in the family Anaplasmataceae that infect leukocytes. Much of the knowledge gained about ehrlichiae originally came from the veterinary sciences, with various studies on *Ehrlichia* (*Cowdria*) *ruminantium* (cattle, sheep, goats) and *Ehrlichia equi* (horses). Canine ehrlichiosis, caused by *Ehrlichia canis*, wiped out 200 to 300 military working dogs during the Vietnam War [36]. The first human case of ehrlichiosis in the United States was a report in March 1986 of a 51-year-old man who had been bitten by a tick in Arkansas and was sick for 5 days before being admitted to a hospital in Detroit [37]. He had malaise, fever, headache, myalgia, pancytopenia, abnormal liver function, renal failure, and high titers of *E. canis* antibodies that fell sharply during convalescence. Physicians assumed he had the dog disease. It turned out not to be the case; he had infection with *E. chaffeensis* (a hitherto unknown agent). For this reason, there are several reports in the medical literature of human infection in the United States with *E. canis*, when, in fact, human ehrlichiosis is usually caused by several closely related *Ehrlichia* organisms. The first one, *Ehrlichia chaffeensis*, is the most frequently reported and is the causative agent of human monocytic ehrlichiosis (HME), which occurs mostly in the southern and south central United States (sporadic cases of HME have also been reported in Europe) and infects mononuclear phagocytes in blood and tissues (Fig. 4.15) [38]. HME is a significant disease—1288 cases were reported to the CDC in 2015 (Fig. 4.16) [9]. The second agent, *E. ewingii*, mostly a dog and deer pathogen, infects granulocytes and causes a clinical illness similar to HME but thus far has only been identified in a few patients, most of whom were immune-compromised. A recent study suggests that cases of *E. ewingii* infection are underreported [39]. The third *Ehrlichia*, the *E. muris*-like agent (EMLA or *Ehrlichia muris eauclairensis*), causes fever, malaise, headache, lymphopenia, and elevated liver enzymes [40]. Thus far, it has only been reported in a couple hundred patients from the upper midwestern United States (mostly Wisconsin), and the vector is the deer tick, *Ixodes scapularis* [41].

**Fig. 4.15** Canine
macrophage with
intracytoplasmic
aggregations (morulae).
(Photo courtesy Dr. Andrea
Varela-Stokes, Mississippi
State University, used with
permission)

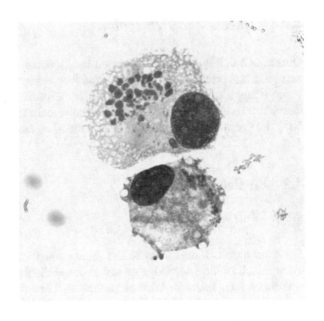

Ehrlichiosis Incidence, 2010

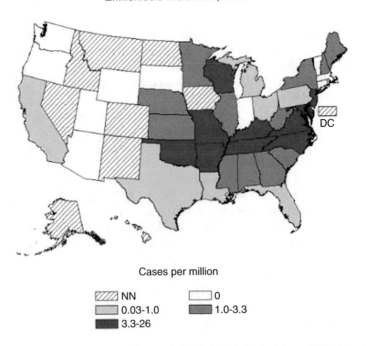

**Fig. 4.16** Geographic distribution of human ehrlichiosis in the United States (CDC figure)

## 4.5.2   Clinical and Laboratory Findings

Clinical and laboratory manifestations of ehrlichial infection are similar [42]. Patients usually present with fever, headache, myalgia, progressive leukopenia (often with a left shift), thrombocytopenia, and anemia. Very often, there are moderate elevations in levels of hepatic transaminases. Sometimes there is a cough, gastroenteritis, or even meningitis. Only about 2–40% of the time is there a rash (more common with HME). Diagnosis depends primarily on clinical findings, although IFA tests may be used to detect antibodies against the respective *ehrlichial* agents. As for serologic tests, the gold standard is IFA performed on paired acute and convalescent sera demonstrating a fourfold rise in antibody titers. Antibodies may not be detectable in the first week of illness, so a negative test during that time does not rule out infection. At one time, human anaplasmosis (see next section) was classified as "human granulocytic ehrlichiosis," and due to this confusion over HGE/HGA terminology, physicians should carefully check what tests they are ordering. Further, due to serological cross-reactivity, tests for ehrlichiosis alone in anaplasmosis-endemic areas may result in an inaccurately high ehrlichiosis incidence and under-recognition of actual anaplasmosis cases [43].

## 4.5.3   Ecology of Ehrlichiosis

Ehrlichiosis is transmitted to humans by tick bites, but different ehrlichial agents have different tick vectors. *Ehrlichia chaffeensis* and *E. ewingii* are both transmitted by the lone star tick (LST) (Fig. 4.17), while the EMLA has only been found so far in *Ixodes scapularis*. LSTs generally occur from central Texas east to the Atlantic Coast and north to approximately Iowa and Maine (Fig. 4.18). WT deer, possibly along with dogs or other mammals, serve as reservoir hosts for the agent, and LSTs are the most important vectors; however, detection of the HME agent in other tick vectors and a few cases outside the distribution of LST indicates that additional vectors occur.

## 4.5.4   Treatment and Control of Ehrlichiosis

Prevention of ehrlichiosis is essentially the same as that of RMSF (*see* Sect. 4.2.4). Dumler and Bakken [38] point out that treatment should be solely a tetracycline, such as doxycycline (possibly rifampin may be an alternate for pregnant patients). Treatment should never be delayed while awaiting laboratory confirmation. Recommended therapy in adults or children is oral or intravenous

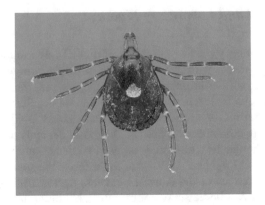

**Fig. 4.17** Adult female lone star tick, *Amblyomma americanum*, an aggressive human biting species. (Photo courtesy Dr. Blake Layton, Mississippi State University)

**Fig. 4.18** Approximate geographic distribution of the lone star tick

doxycycline [10, 12]. Fever typically subsides within 24–48 h after treatment when the patient receives doxycycline or another tetracycline during the first 4–5 d of illness. If a patient fails to respond to early treatment, this response might be an indication that their condition is not a tick-borne rickettsial disease.

## 4.6  Anaplasmosis

### 4.6.1  Introduction

*Anaplasma phagocytophilum* infects granulocytes and causes human granulocytic anaplasmosis (HGA). There is confusion about this terminology because for many years, this disease was called human granulocytic ehrlichiosis (HGE) and is often included in the older medical literature under that label. Complicating matters further, sometimes commercial laboratories may still refer to tests for HGA as human granulocytic ehrlichiosis tests. HGA mostly occurs in the upper midwestern and northeastern United States. There were 3656 cases of HGA reported to the CDC in 2015, a 30% increase over 2014 [9]. The case fatality rate is 0.3% but may be higher in older patients [44].

### 4.6.2  Clinical and Laboratory Findings

Clinical and laboratory manifestations of HGA infection include fever, headache, myalgia, progressive leukopenia (with a left shift), thrombocytopenia, and anemia. Often there are moderate elevations in levels of hepatic transaminases. Sometimes there is a cough, gastroenteritis, or meningitis. Illness due to HGA is somewhat milder than with HME; reported fatality rates are about 1% and 2% for HGA and HME, respectively. Research indicates that both agents alter the patient's immune system, allowing opportunistic infections such as fungal pneumonia to occur. Diagnosis is made on clinical findings, although IFA tests may detect antibodies against the anaplasmal agent. The gold standard serologic test for anaplasmosis is IFA performed on paired acute and convalescent sera demonstrating a fourfold rise in antibody titers. As is the case with ehrlichiosis, antibodies may not be detectable in the first week of illness, so a negative test during that time does not rule out infection. Due to confusion over HGE/HGA terminology, physicians should carefully check what tests they are ordering. If HGA is suspected, physicians should make sure the test they order detects antibodies to *Anaplasma phagocytophilum*.

**Fig. 4.19** Adult female deer tick, *I. scapularis.* (Photo courtesy Dr. Blake Layton, Mississippi State University)

### 4.6.3   Ecology of Anaplasmosis

Anaplasmosis (HGA) is transmitted to humans by the deer tick, *Ixodes scapularis* (Figs. 4.19 and 4.20). The ecology of HGA is not well known. It has been diagnosed mostly in patients from the upper Midwest, the northeastern United States, and the Pacific Coast area, although cases have also been reported sporadically elsewhere, including Europe and Asia. The tick vector in the United States is *Ixodes scapularis* which is the same species that transmits the agent of Lyme borreliosis; thus, there is the possibility of coinfection with Lyme borreliosis and HGA (and even babesiosis) [45]. In Europe and Asia, the vectors of anaplasmosis are *I. ricinus* and *I. persulcatus*, respectively. Animal reservoirs of the HGA agent may be a variety of small rodents and possibly deer.

### 4.6.4   Treatment and Control of Anaplasmosis

Prevention of anaplasmosis is essentially the same as that of other tick-borne diseases such as RMSF (*see* Appendix 3). Treatment should never be delayed while awaiting laboratory confirmation. Recommended therapy in adults or children is oral or intravenous doxycycline [10, 12, 46]. There is some evidence that treatment of children and pregnant women with rifampin is successful [46].

**Fig. 4.20** Approximate geographic distribution of the deer tick

**What's Going on with Lyme Disease in the South?**
There is controversy about whether or not true Lyme disease (LD) occurs in the southern United States. Some physicians and researchers are convinced that it does, and numerous cases are reported to state health departments and the CDC each year. In fact, Mississippi (my state) receives reports of about 30 cases of LD annually. Evidence of LD in the South, proponents say, includes clinical syndromes consistent with LD, serologic test results sometimes indicative of infection with *Borrelia burgdorferi*, and rashes that resemble the EM lesion. Other physicians adamantly contend that there is no LD in the South. They show as evidence invariably negative data from extensive retesting and

(continued)

follow-up of patients with suspected Lyme disease. For example, in 1999, the Mississippi Department of Health investigated 48 cases of physician-diagnosed, locally acquired LD. Each medical record was reviewed, and blood samples were drawn for enzyme-linked immunosorbent assay (ELISA) and Western blot analysis. Results indicated that only one sample was ELISA-positive; none were positive by Western blot (S. Slavinski, MD, Mississippi Department of Health, personal communication, August 2003). Therefore, no evidence of infection with *B. burgdorferi* could be found.

Cases of a Lyme disease-like illness—cases that meet the CDC case definition for Lyme disease—do occur in the South. Interestingly, these cases may respond to treatment with antibiotics, suggesting a bacterial cause of some type. However, it's not clear whether these cases would have resolved on their own without antibiotics. A bona fide (widely accepted) human isolate from a patient in this part of the country is lacking despite numerous attempts to isolate organisms from EM lesions. A small percentage of lone star ticks, *Amblyomma americanum*, harbor spirochetes that react with reagents prepared against *B. burgdorferi*. These spirochetes are a true *Borrelia* species which has been tentatively named *Borrelia lonestari*. For several years, scientists thought that some Lyme disease-like illnesses in the southern United States were caused by this new spirochete, but now evidence is leaning away from *B. lonestari* as the etiologic agent. Perhaps other, as of yet undescribed, microorganisms are involved.

There is no reason why LD should not occur in the southern United States. The tick vector, *Ixodes scapularis*, is found in the South. And there have been isolations of true *B. burgdorferi* from both rodents and ticks in the southern states (South Carolina, Florida, Georgia, and Texas). However, this is very rare. Contrast this with the northeastern United States, where about 50% of *I. scapularis* ticks are infected. Some of this infection disparity can be explained by the fact that southern *I. scapularis* nymphs prefer to feed on lizards and skinks which are incompetent reservoirs for *B. burgdorferi*.

Further complicating the issue is an apparent hypersensitivity reaction to saliva of the lone star tick that sometimes occurs 1–3 d following a bite. This hypersensitivity reaction resembles EM and is often 6–8 cm in diameter, ring-like, raised, and vesicular Fig. 4.24). While studies of such lesions are lacking, they are probably not true EM lesions because there is little or no incubation period, the lesions often fade in a few days, and the lesions are raised (vesicular). EM lesions *may* be vesicular but usually are not. In fact, they are often flat, almost imperceptible by touch. In southern states where physicians do not see many cases of true LD, these hypersensitivity reactions may be misdiagnosed as the real thing.

While some researchers insist that LD does not occur in the southern United States, it is unwise at this point to exclude LD from the differential diagnosis in persons with a possible tick-borne illness in this region. Empirical evidence supports the presence of a Lyme-like illness in the South, perhaps caused by other organisms.

## 4.7  Lyme Disease

### 4.7.1  Introduction

Lyme disease (LD) is a systemic tick-borne illness which may present with many clinical manifestations such as cardiac, neurologic, and joint problems. In the United States, LD is caused by the bacterium, *Borrelia burgdorferi* sensu stricto, although another agent *B. mayonii* has been shown to cause human illness [47]. In Europe, there are several other *Borrelia* species involved. Initial symptoms of LD include a flulike syndrome with headache, stiff neck, myalgias, arthralgias, malaise, and low-grade fever. Often, a more or less circular, painless, macular dermatitis is present at the bite site called erythema migrans (EM). The EM lesion is sometimes said to be pathognomonic for LD, although not all patients develop it. EM lesions may steadily increase in size with or without subsequent central clearing. Numbers of reported LD cases in Europe may be in the hundreds of thousands [48], while in the United States, officially there were 38,069 cases reported by the CDC in 2015 [9], although an indirect estimate is 300,000 per year [49]. Although cases occur in most states and the District of Columbia, the vast majority are from the northeastern and north central United States (Fig. 4.21). In fact, between 2008 and 2015, 14 states in the northeast, mid-Atlantic, and upper midwest regions accounted for 95.2% of all reported cases, and 95.7% of all confirmed cases reported in the United States [50].

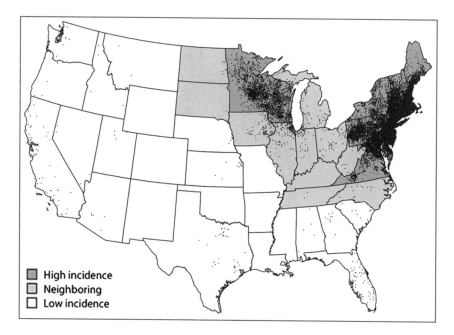

**Fig. 4.21** Lyme disease case distribution, United States (CDC figure)

### 4.7.2  Background and Historical Information

Lyme disease is named after the town it was first described from—Old Lyme, Connecticut (Fig. 4.22). Unlike some diseases named after people (e.g., Chagas' disease, Hansen's disease), LD is named after a *place*; therefore, the term "Lyme's disease" is incorrect. The story of the recognition of LD in the United States is fascinating. In 1975, there was a geographic clustering of children near Lyme Connecticut who were ill with what appeared to be juvenile rheumatoid arthritis. This prompted researchers from Yale University School of Medicine to investigate this "new" disease [51]. Clues to the infectious nature of the disease included clustering of cases, history of tick bite in most patients, and attenuation of the arthritis by antibiotic therapy. Soon the search was on for the causative agent of the disease. A major breakthrough came in 1982. Serendipitously, Willy Burgdorfer, a medical entomologist working at the Rocky Mountain Laboratories in Hamilton, Montana, found spirochetes in the midgut of ticks sent to him from Shelter Island, New York—a place known to have endemic LD. He rightly assumed that he had found the sought after causative agent of LD and proceeded to culture the organism. In subsequent experiments using the newly found spirochetes and infected ticks, he demonstrated EM lesions on white rabbits 10–12 wk. after infected ticks had fed on them. Other experiments showed that patient serum

**Fig. 4.22**  Road sign from Old Lyme Connecticut, the place where Lyme disease was first described in the United States. (Photo courtesy Pat and Tom Baker)

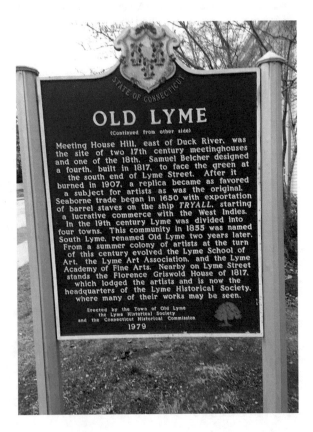

reacted strongly to the newly isolated spirochete. Thus, Burgdorfer and colleagues are credited with finding the causative agent of LD [8]. The spirochete was subsequently named *Borrelia burgdorferi* in honor of Burgdorfer. Later, additional conclusive evidence was produced when Steere et al. [52] isolated the same spirochete out of blood and EM skin lesions of patients with LD.

The historical background of the main tick vector of LD is filled with controversy, not controversy about the tick's identification, but, instead, what to call it. There has been a deer tick, *Ixodes scapularis*, known to occur in the southern United States since the 1800s (Fig. 4.19). During the early days of the investigation of LD, a similar tick was described from the outbreak area in the northeastern United States. It was morphologically very similar to *I. scapularis* but was thought to differ enough to warrant species status. Thus, this "new" species was named *Ixodes dammini* [53], but shortly thereafter, tick taxonomists begin to recognize that although the two species were different at the extremes of their distributions, in the intergrading zones— where the two species met—they were extremely difficult to separate. Suggestions soon circulated among tick specialists that the two species were one and the same. Oliver et al. [54] addressed this issue, producing evidence showing mating compatibility and genetic similarity, and thus claimed that the two were definitely one species. Under rules set forth by the International Commission on Zoological Nomenclature, the older name, *I. scapularis*, takes precedence. The name *I. dammini* goes away. However, even today some members of the scientific community still try to resurrect the old name. This author considers the expertise of the former world's foremost tick taxonomist—Dr. James Keirans—to be authoritative in this matter, and he considered them one species, *I. scapularis* [55].

### 4.7.3 Clinical and Laboratory Findings

The 2008 LD case definition says that a confirmed case is defined as either (1) a person with erythema migrans (EM) with possible exposure to tick habitat in an area where LD is endemic or who has laboratory evidence of infection or (2) a person with at least one other defined clinical manifestation of LD and (also) laboratory evidence of infection. As mentioned earlier, EM is a skin lesion that typically begins as a red macule or papule and expands over a period of days or weeks to form a large round lesion, sometimes with partial central clearing (Fig. 4.23). The case definition requires that EMs be at least 5 cm in diameter. However, smaller and atypical lesions may occasionally occur. Not all EM lesions are due to LD [56], and all such lesions must be distinguished from other manifestations, such as strep and staph cellulitis, hypersensitivity reactions to tick bite, plant dermatitis, various fungal infections, and granuloma annulare [57]. EM occurs in 60–80% of LD cases and is often accompanied by mild to moderate constitutional symptoms such as fatigue, fever, headache, mild stiff neck, arthralgias, myalgias, and regional adenopathy. Untreated EM and associated symptoms usually resolve in 3–4 wk. However, the disease often disseminates within weeks or months, resulting in cardiac, neurologic, and joint manifestations. Symptoms may include Lyme carditis, cranial neuropathy, radiculopathy, diffuse peripheral neuropathy, meningitis, and asymmetric oligoarticular arthritis.

**Fig. 4.23** Erythema
migrans lesion from Lyme
disease (CDC photo)

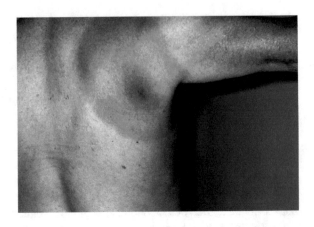

Immunoglobulin M (IgM) antibody generally first develops within 2–4 wk after the appearance of EM, peaks after 6–8 wk of illness, and declines to normal range after 4–6 mo of illness in most patients [58]. IgG levels are usually elevated within 6–8 wk after onset of LD. Laboratory evidence of infection with *B. burgdorferi* is usually established when a laboratory:

1. Isolates the spirochete from tissues or body fluids
2. Detects positive two-tier IgM or IgG antibodies in an ELISA test followed by reflex immunoblot (This has to be properly interpreted using established criteria.)

In addition, some labs and university medical centers have capability of detecting *B. burgdorferi* DNA by PCR. PCR is a very useful laboratory tool. Body fluids from patients, such as blood, urine, CSF, and synovial fluid, are good candidates for PCR analysis. However, laboratories conducting PCR may become contaminated, leading to false positives. Perhaps the best method of confirming infection with LD at this time is detection of IgM and IgG antibodies with ELISA tests, followed up with Western blot analysis. Western blotting is valuable in distinguishing true-positive from false-positive ELISA results. Lab analysis is complicated by the fact that persons who lack antibodies to *B. burgdorferi* during early weeks of infection may not develop antibodies following antibiotic therapy [59]. On the other hand, patients who have antibodies to *B. burgdorferi* and who are effectively treated and cured may continue to carry these immunoglobulins for several months or years.

### 4.7.4   Ecology of LD

LD is solely tick-borne. In the United States, *I. scapularis* is the primary vector in the East and *Ixodes pacificus* in the West. Each of the three motile life stages of hard ticks must get on a host, feed, fall off, and then transform into the next stage. If no blood-providing host is available, the ticks will perish. Therefore, an important aspect of vector-borne disease ecology is host availability, and not just availability, but diversity as well. If immature ticks feed on hosts that are refractory to infection with the LD spirochete, then overall prevalence of the disease agent in an area will decline. On the other hand, if an abundant host is available that also is able to be infected with *B. burgdorferi* producing long and persistent spirochetemias, then prevalence of tick infection increases. This is precisely the case in the northeastern and upper midwestern states. In those areas, the primary host for immature *I. scapularis* is the white-footed (WF) mouse, which is capable of infecting nearly 100% of larval ticks during feeding. Since infection can be transferred from tick stage to tick stage, this obviously leads to high numbers of infected nymphs and adults. In the West and South, tick infection rates are much lower (and hence, lower numbers of LD cases). This is attributed to the fact that immature stages of *I. scapularis* and *I. pacificus* feed primarily on lizards, which are incompetent as reservoirs and incapable of infecting ticks. Another factor affecting the dynamics of LD is the fact that nymphal *I. scapularis* are the stage primarily biting people and transmitting the disease agent in the Northeast, whereas in the South, nymphal *I. scapularis* rarely, if ever, bite humans. In fact, they are very difficult to find even in areas known to have them [60]. Adult ticks are certainly capable of transmitting the LD agent in all areas—North, South, or West—but adult ticks are large enough to be easily seen and removed by people. Nymphs, on the other hand, are about the size of the head of a pin and may be easily overlooked or confused with a freckle.

Other tick species may be involved in the ecology of LD. In the southern United States, there have been reports for years about an LD-like illness [61], which other researchers have voiced doubts about—doubts regarding whether or not it is true LD. In fact, the CDC often labels these southern Lyme-like illnesses as southern tick-associated rash illness (STARI) or Master's disease (Fig. 4.24). Cases of STARI may be due to allergic reactions to tick saliva or other (as yet) unknown causes [62].

## 4.7.5  Treatment

Early LD responds readily to oral antibiotics, such as doxycycline, amoxicillin, cefuroxime, or azithromycin, which are generally prescribed for 2–3 wk [10, 13, 57]. The duration of antibiotic administration should be individualized according to the severity of illness and the rapidity of clinical response. Children younger than 9 can be treated with amoxicillin [10]. Late-stage LD is a controversial diagnosis and may be more difficult to treat. Consultation with an infectious disease physician is recommended in those cases. Deciding who to treat for LD is frequently a problem since early LD is diagnosed by clinical presentation alone. Also, as mentioned above, the disease's most recognized sign (EM) may be confused with other skin lesions. Finally, in many bona fide cases of LD, patients are initially seronegative and will remain so if antibiotic treatment is begun early.

**Fig. 4.24** STARI rash due to bite of the lone star tick, 10 d post-bite. (Photo copyright 2011 by Jerome Goddard, Ph.D.)

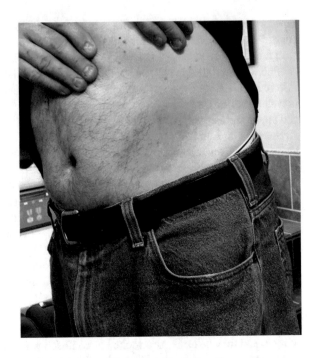

## 4.8  Tularemia

### 4.8.1  Introduction and Medical Significance of Tularemia

Tularemia, sometimes called rabbit fever or deer fly fever, is a bacterial zoonosis that occurs throughout temperate climates of the northern hemisphere. Historically, approximately 150–300 cases have occurred in the United States each year, with most cases occurring in Arkansas, Missouri, and Oklahoma [13]. There were 314 cases of tularemia reported in 2015, the highest number since 1964 [9]. The causative organism, *Francisella tularensis*, is a highly infectious small, Gram-negative, nonmotile cocco-bacillus named after Sir Edward Francis (who did the classical early studies on the organism) and Tulare, California (where it was first isolated). The disease may be contracted in a variety of ways—food, water, mud, articles of clothing, contact with infected animal tissue, and (particularly) arthropod bites. Arthropods involved in transmission of tularemia include ticks, biting flies, and possibly even mosquitoes (Fig. 4.25). Ticks account for more than 50% of all cases, especially west of the Mississippi River. There are four subspecies of tularemia organisms [63]. Two of them are primarily associated with human disease, namely, *F. tularensis*, subspecies *holarctica* (Jellison type B), and *F. tularensis*, subspecies *tularensis* (Jellison type A) [64]. Type A is the most virulent and is present only in North America. Tularemia may present as several different clinical syndromes, including glandular, ulceroglandular, oculoglandular, oropharyngeal, pneumonic, and typhoidal [65]. In general, the clinical course is characterized by an

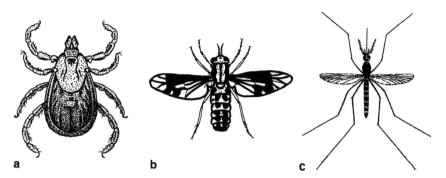

Fig. 4.25 Some arthropods reported to be involved in transmission of tularemia to humans: (**a**) tick, (**b**) deer fly, (**c**) mosquito. (Provided with permission by *Infections in Medicine* 1998;15:306)

influenza-like attack with severe initial fever, temporary remission, and a subsequent febrile period of at least 2 wk. Later, a local lesion with or without glandular involvement may occur. Additional symptoms vary depending on the method of transmission and form of the disease (*see* discussion below). Untreated, the mortality rate for tularemia is as high as 30%; early diagnosis and treatment can reduce that 5% or less [10, 66].

## 4.8.2   Clinical and Laboratory Findings

Depending on the route of entry of the causative organism, tularemia may be classified in several ways. The most common is ulceroglandular—resulting from cutaneous inoculation—characterized by an ulcer with sharp undermined borders and a flat base. Location of the ulcers may help identify the mode of transmission. Ulcers on the upper extremities are often a result of exposure to infected animals, whereas ulcers on the lower extremities, back, or abdomen most often reflect arthropod transmission. When there is lymphadenopathy without an ulcerative lesion, the classification of glandular tularemia is used. If the tularemia bacterium enters via the conjunctivae, oculoglandular tularemia may result. Oropharyngeal tularemia results from ingestion of contaminated food or water. If airborne transmission of the agent is involved, the pneumonic form occurs. These patients often present with fever, a nonproductive cough, dyspnea, and chest pain. The pneumonic form can be life-threatening. Finally, tularemia may be classified as typhoidal, characterized by disseminated infection mimicking typhoid fever, brucellosis, tuberculosis, or some of the RMSF-type infections.

Patients with tularemia may or may not show abnormal white blood cell counts (WBC), platelet counts, and sedimentation rate. Hyponatremia, elevated serum transaminases, increased creatine phosphokinase, and myoglobinuria have been reported [66]. The standard serologic test used to confirm tularemia has historically involved tube agglutination of a bacterial suspension. The test is quite specific for tularemia, although it can cross-react with *Brucella*. An acute agglutination titer of 1:160 is supportive of a tularemia diagnosis, but definitive evidence of recent infection comes from a fourfold rise in titer between acute and convalescent specimens. In recent years, diagnosis has been aided by detecting *Francisella tularensis* DNA in blood or tissues by PCR.

## 4.8.3   *Arthropod Transmission of the Tularemia Organism*

Original work by Francis in the 1920s established that the plague-like disease of rodents in California and Utah was caused by *Bacterium tularense* (the name was later changed to *F. tularensis*), transmitted by deer flies [67, 68]. Later, Parker et al. reported finding the organism in the Rocky Mountain wood tick, *D. andersoni* [69]. By the 1930s, tularemia organisms had been found naturally occurring in several tick species, and new information was acquired about the animal hosts and various methods of tularemia transmission. Tularemia was turning out to be a complex zoonosis.

In subsequent field studies, many animals were found with evidence of tularemia infection, but determining the actual reservoirs was not an easy task. Hopla [70] detailed the history of tularemia and summarized field and laboratory data on its ecology. Numerous animal species are susceptible to tularemia (at least 47 species of mammals and birds), but especially the cottontail rabbit [70]. Many cases of tularemia result from skinning rabbits during hunting season. In fact, two peaks of tularemia case numbers may often be seen in the southern states—one in the summer resulting from tick bites and the other in the fall resulting from skinning rabbits. Dogs and cats are not thought to be reservoirs of *F. tularensis* or essential for maintenance of the organism in an ecosystem. However, they may be disseminators of the tick vectors (particularly the American dog tick), bringing ticks into close contact with humans. It is also possible for cats to transmit the bacteria via the claws after killing or feeding on infected prey [71].

Tularemia infection in ticks occurs in both the gut and body tissues and hemolymph fluid (tick blood). Infection is known to persist for many months and even years in some species. Tularemia organisms may be passed from tick stage to tick stage and to the offspring of infected female ticks. The three major North American ticks involved in transmission of tularemia organisms are the LST, *A. americanum* (Fig. 4.17); the Rocky Mountain wood tick, *D. andersoni* (Fig. 4.6); and the American dog tick, *D. variabilis* (Fig. 4.8). Both the LST and the American dog tick occur over much of the eastern United States; the Rocky Mountain wood tick occurs in the West. All three of these tick species are avid human biters. In fact, the LST is so numerous in the southern and south central United States that almost every person who goes outdoors gets bitten by one or more stages of this tick.

**Fig. 4.26** *I. ricinus*, vector
of LD, babesiosis, TBE,
and tularemia organisms in
Western and Central
Europe (Specimen
courtesy Dr. Lorenza Beati,
U.S. National Tick
Collection)

1 mm

However, not all ticks (even within a vector species) are infected—generally only
a very small percentage. In Central and Western Europe, *Ixodes ricinus* is probably a vector (Fig. 4.26). *D. nuttalli* may be a vector in Russia.

Other arthropods are involved in the transmission of tularemia organisms as
well, though not to the extent that ticks are. Some species of deer and horse flies
are proven vectors (Fig. 4.25b). In fact, some of the original work by Francis and
Mayne demonstrated transmission by the deer fly, *Chrysops discalis* [67].
Mosquitoes may be involved in transmission, but information is scant at this time.
Extreme caution should be exercised in making interpretations whenever tularemia organisms are isolated from insects. It is fairly common for a blood-feeding
arthropod to "pick up" microbial pathogens with their blood meal. Isolation of an
etiologic agent from a blood-feeding arthropod in no way implies that the arthropod is an effective vector of that agent. Myriad factors influence vector competence (i.e., ability or inability to pick up a pathogen and later transmit it to another
host) (*see* Chap. 2).

### 4.8.4   Treatment

The drug of choice for treatment of tularemia is streptomycin, although the recent
lack of availability of this drug has forced many healthcare providers to try alternative antimicrobials, such as gentamicin, tetracycline, chloramphenicol, and others

[71, 72]. Unfortunately, controlled studies are lacking to support the efficacy of some of these products, and some agents are only inhibitory—not bactericidal—thus leading to relapses. A review of the literature found the following cure rate data for some of the most effective agents: streptomycin, 97%; gentamicin, 86%; tetracycline, 88%; and chloramphenicol, 77% [72]. Tetracycline was shown to be associated with twice as many relapses as gentamicin. The authors of that study concluded that gentamicin was comparable to streptomycin in efficacy against tularemia [71].

## 4.9   Human Babesiosis

### 4.9.1   Introduction and Medical Significance

Human babesiosis is a tick-borne disease primarily caused by protozoa in the order Piroplasmida, family Babesiidae. The most common human agents are *Babesia microti* and *Babesia divergens*, although other newly recognized species may also cause human infection [73, 74]. The disease is a malaria-like syndrome characterized by fever, fatigue, and hemolytic anemia lasting from several days to a few months. In terms of clinical manifestations, babesiosis may vary widely from asymptomatic infection to a severe, rapidly fatal disease. The first demonstrated case of human babesiosis in the world was reported in Europe in 1957 [75]. Since then, there have been at least 40 additional cases in Europe. Most European cases occurred in asplenic individuals and were caused by *Babesia divergens*, a cattle parasite. In the United States, there have been hundreds of cases of babesiosis (most people with intact spleens) caused by *Babesia microti*, mostly from southern New England and specifically Nantucket, Martha's Vineyard, Shelter Island, Long Island, and Connecticut [76, 77]. The tick vector in Europe is believed to be the European castor bean tick, *I. ricinus* (Fig. 4.26), one of the most commonly encountered ticks in central and western Europe. In the United States, most cases of babesiosis are caused by bites from the same tick that transmits the agent of Lyme disease, *I. scapularis* [78] (Figs. 4.19 and 4.20).

### 4.9.2   Clinical and Laboratory Findings

Babesiosis is clinically very similar to malaria; in fact, confusion between the two diseases is often reported in the scientific literature [79]. Headache, fever, chills, nausea, vomiting, myalgia, altered mental status, disseminated intravascular coagulation, anemia with dyserythropoiesis, hypotension, respiratory distress, and renal insufficiency are common to both diseases. However, the symptoms of babesiosis do not show periodicity. The incubation period varies from 1 to 4 wk. Physical exam of patients is generally unremarkable, although the spleen and liver may be palpable. Lab findings may include hemoglobinuria, anemia, and elevated serum

bilirubin and transaminase levels. Diagnosis of babesiosis is usually based on recognition of the organism within erythrocytes in Giemsa-stained blood smears,
although PCR with specific *Babesia* primers is much more sensitive. The small
parasites, several of which may infect a single red blood cell and which appear
much like *Plasmodium falciparum*, can be differentiated from malarial parasites by
the absence of pigment (hemozoin) in the infected erythrocytes. Laboratory animals
may be useful in diagnosis of babesiosis. Specialized laboratories have the capability to inject patient blood into hamsters and subsequently detect parasitemias 2–4
wk after inoculation. IFA can be used to detect specific antibodies in patient serum.
Serologic diagnosis can be established by a fourfold or greater rise in the serum titer
between the acute phase and the convalescent phase.

## 4.9.3  Species of Babesia and Their Ecology

Babesial parasites, along with members of the genus *Theileria*, are called piro-
plasms because of their pear-shaped intraerythrocytic stages. There are at least 100
species of tick-transmitted *Babesia*, parasitizing a wide variety of vertebrate animals. Some notorious ones are as follows: *Babesia bigemina*, the causative agent of
Texas cattle fever; *Babesia canis* and *Babesia gibsoni*, canine pathogens; *Babesia
equi*, a horse pathogen that occasionally infects humans; *Babesia divergens*, a cattle
parasite that infects humans; and *Babesia microti*, a rodent parasite that infects
humans. Recently, new *Babesia* species have been recovered from ill humans and
have tentatively been variously designated as the WA1 agent, the CA1 agent, or the
MO1 agent [80]. The WA1 agent, now known as *B. duncani*, was isolated from a
patient in Washington State and was particularly interesting because the man was
only 41 yr old, had an intact spleen, and was immunocompetent. Although the parasites were morphologically identical to *B. microti*, the patient did not develop a
substantial antibody to *B. microti* antigens. Subsequent DNA sequencing of the
organism indicated that it was most closely related to the canine pathogen *B. gibsoni*. The probable tick vector for the WA1 agent is the western black-legged tick, *I.
pacificus*. Obviously, there is much more to be learned about the many and varied
*Babesia* species and their complex interactions with animals in nature.

On the other hand, the life cycle of *B. microti*—the one causing American babesiosis in the northeastern United States—is fairly well known. Rodents serve as
natural reservoirs for the parasite. *B. microti* multiplies readily in hamsters and the
white-footed (WF) mouse, *Peromyscus leucopus*. In fact, the WF mouse is the preferred natural host for *B. microti* and is also the host for immature *I. scapularis*
ticks. Immature stages of the ticks "pick up" the parasites in their blood meal from
the rodents and subsequently transmit them (in a later tick stage) to a vertebrate
host, but factors affecting the host–vector–pathogen relationship are ever-changing.
WF mouse populations are cyclic, depending on food sources, and are more abundant some years compared to others. If ticks happen to feed during those times on
an animal (such as a squirrel) that is somewhat refractory to infection with *B.*

*microti*, then the diversion from reservoir-competent hosts depresses the overall infection rate in ticks and mice. Deer play a role as well—but not as reservoirs—in providing a blood meal for the adult *I. scapularis*. More deer ultimately lead to more ticks. Therefore, prevalence of *B. microti* infection in an area depends on the complex interactions of WF mice, the parasite, and deer.

### 4.9.4  Treatment and Control

Standard treatment of symptomatic *B. microti* infection has been quinine sulfate plus clindamycin; however, a drug regimen consisting of atovaquone and azithromycin has been shown to be effective when clindamycin and quinine fail [10, 81]. Prevention and control of the disease in the community involve personal protection measures against ticks (*see* Appendix 3), searching for and removing promptly any attached ticks, pesticidal treatment of lawns and parks to reduce tick numbers, and possibly host animal management (deer reduction).

## 4.10  Viruses Transmitted by Ticks

### 4.10.1  Introduction

People usually associate arboviral encephalitis and dengue-like fevers with mosquitoes. However, ticks may also be involved in the transmission of these types of agents. Tick-borne viral diseases (nonhemorrhagic) have historically been grouped into two categories—the encephalitides group and the dengue fever-like group. The former, containing viral diseases clinically resembling the mosquito-borne encephalitides, includes the tick-borne encephalitis (TBE) subgroup (and various subtypes). Specific diseases in this subgroup have historically included Central European TBE, Russian spring-summer encephalitis (RSSE), Louping ill, and Powassan encephalitis (POW). The virus species have been variously renamed and regrouped as Far Eastern (previously RSSE), Siberian (previously West-Siberian), and Western European (previously Central European encephalitis) [82]. Historically, POW has been the only one of these occurring in North America. (Note: The viruses of Omsk hemorrhagic fever and Kyasanur forest disease are in the TBE complex but produce hemorrhagic fevers and also differ in many other epidemiologic and ecologic features. Therefore, they will not be discussed here.) The major dengue-like viral disease transmitted by ticks is Colorado tick fever (CTF).

In the last decade, new tick-borne viruses have been identified. Heartland virus (a *Phlebovirus*) is associated with the lone star tick, *Amblyomma americanum*, and has been recognized in Missouri, Oklahoma, Kentucky, and Tennessee [83, 84]. Only about 30 cases of Heartland virus have been identified. A couple of cases of a new *Thogotovirus* called Bourbon virus have been identified in the Midwest and southern United States with an unknown tick vector [85]. Recent evidence suggests the lone star tick may be a potential vector [86].

## 4.10.2 Tick-Borne Encephalitis (TBE)

### 4.10.2.1 Clinical and Epidemiologic Features

TBE should be considered a general term encompassing several diseases caused by similar flaviviruses spanning from the British Isles (Louping ill), across Europe (Central European TBE), to the Far East (RSSE and similar syndromes). These diseases also differ in severity—Louping ill being the mildest and Far-Eastern form (RSSE) being the worst. In Central Europe, the typical case has a biphasic course with an early, viremic, flulike stage, followed about a week later by the appearance of signs of meningoencephalitis [87]. Central nervous system (CNS) disease is relatively mild, but occasional severe motor dysfunction and permanent disability occur. The case fatality rate is 1–5% [88]. RSSE (sometimes referred to as the Far-Eastern form) is characterized by violent headache, high fever, nausea, and vomiting. Delirium, coma, paralysis, and death may follow; the mortality rate is about 25–30% [89]. Louping ill—named after a Scottish sheep disease—in humans also displays a biphasic pattern and is generally mild. As mentioned, the virus infects sheep; few cases are actually ever reported in humans. Reported case numbers for TBE are estimated to be as many as 14,000 per year [90]. Transmission to humans is mostly by the bite of an infected tick. However, infection may also be acquired via consuming infected milk and uncooked milk products. The distribution and seasonal incidence of TBE are closely related to the activity of the tick vectors—*I. ricinus* in western and central Europe (Fig. 4.24) and *Ixodes persulcatus* in central and eastern Europe (there is overlap of the two species). *I. ricinus* is most active in spring and autumn. Two peaks of activity may be observed: one in late March to early June and one from August to October. *I. persulcatus* is usually active in spring and early summer. Apparently, *I. persulcatus* is more cold hardy than *I. ricinus*, thus inhabiting harsher, more northern areas.

Powassan encephalitis (POW), also part of the TBE serocomplex, is a relatively rare infection of humans that mostly occurs in the northeastern United States, adjacent regions of Canada, and parts of Russia. There were seven cases reported in the United States during 2015 [9]. Cases are characterized by sudden onset of fever with temperature up to 40 °C along with convulsions. Also, accompanying encephalitis is usually severe, characterized by vomiting, respiratory distress, and prolonged, sustained fever. Only a few dozen cases of POW have been reported in North America [91, 92], although its incidence may be increasing [9]. Cases have occurred in children and adults, with a case fatality rate of approximately 50%. POW is maintained in an enzootic cycle among ticks (primarily *Ixodes cookei*) and rodents and carnivores. *Ixodes cookei* only occasionally bites people, and this may explain the low case numbers. Antibody prevalence to POW in residents of affected areas is generally <1%, indicating that human exposure to the virus is rare.

Deer tick encephalitis, closely related to POW, is another member of the TBE complex which was first discovered in North America in the late 1990s [93]. There have only been a few clinical cases ever described, although at least one death has been attributed to this virus [94]. The agent has been found along the Atlantic Coast and in Wisconsin and is primarily found infecting the deer tick, *Ixodes scapularis*.

### 4.10.2.2   Diagnosis and Treatment

Definitive diagnosis of TBE is based on isolating the virus from blood or CSF or from postmortem tissues, by PCR, or serologic tests of paired sera, or demonstration of specific IgM in acute serum. Virus isolation is generally an option only at major research hospitals or government institutions. Hemagglutination inhibition is often used to detect antibody rises between early and late serum samples. Enzyme-linked immunosorbent assay (ELISA) tests are used to indicate presence of specific IgM. Treatment is supportive only; no specific treatment is available. A vaccine for TBE (FSME-ImmunInject® Baxter-Immuno, Vienna, Austria) has been shown to be safe and effective through 30 yr of routine use in central Europe [95, 96].

## 4.10.3   Colorado Tick Fever (CTF)

### 4.10.3.1   Clinical and Epidemiologic Features

CTF is a generally moderate, acute, self-limiting, febrile illness caused by a *Coltivirus* in the Reoviridae. Typically, onset of CTF is sudden, with chilly sensations, high fever, headache, photophobia, mild conjunctivitis, lethargy, myalgias, and arthralgias. The temperature pattern may be biphasic, with a 2–3-d febrile period, a remission lasting 1–2 d, and then another 2–3 d of fever, sometimes with worse symptoms [97]. Rarely, the disease may become severe in children with encephalitis, myocarditis, or tendency to bleed. Infrequently, a transient rash may accompany infection. Recovery is usually prompt, but a few fatal cases have been reported. CTF occurs in areas above 4000 feet in at least 11 western states (South Dakota, Montana, Wyoming, Colorado, New Mexico, Utah, Idaho, Nevada, Washington, Oregon, and California) and in British Columbia and Alberta, Canada. Exact case numbers are hard to ascertain because the disease is not nationally notifiable and many cases are so mild that ill persons fail to seek medical care, but about 20 cases are reported in the United States annually (other estimates are 200–400 cases annually). Peak incidence is during April and May at

## Ecology of Colorado Tick Fever Virus

Colorado tick fever (CTF) virus is spread by Rocky Mountain wood ticks (*Dermacentor andersoni*). Rocky mountain wood ticks are found in the western United States and Canada at 4,000–10,000 feet above sea level. Here are the steps in how the virus is spread:

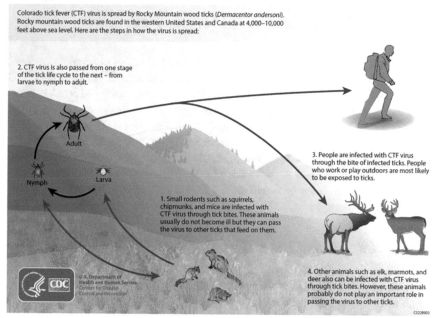

2. CTF virus is also passed from one stage of the tick life cycle to the next – from larvae to nymph to adult.

Adult

Nymph

Larva

1. Small rodents such as squirrels, chipmunks, and mice are infected with CTF virus through tick bites. These animals usually do not become ill but they can pass the virus to other ticks that feed on them.

3. People are infected with CTF virus through the bite of infected ticks. People who work or play outdoors are most likely to be exposed to ticks.

4. Other animals such as elk, marmots, and deer also can be infected with CTF virus through tick bites. However, these animals probably do not play an important role in passing the virus to other ticks.

U.S. Department of Health and Human Services Centers for Disease Control and Prevention

CS228003

**Fig. 4.27**   Colorado tick fever life cycle (CDC figure)

lower elevations and during June and July at higher elevations. The virus is maintained in nature by cycles of infection among various small mammals and the ticks that parasitize them (Fig. 4.27). Infection in humans is by the bite of an infected tick. Several tick species have been found infected with the virus, but *D. andersoni* (Fig. 4.6) is by far the most common. *D. andersoni* is an avid human biter occurring in the Rocky Mountain region of the United States and Canada. It is especially prevalent where there is brushy vegetation to provide good protection for small mammalian hosts of immature ticks and yet with sufficient forage to attract large hosts required for the adults.

### 4.10.3.2   Diagnosis and Treatment

CTF can be confirmed by isolating the virus from blood by inoculation of suckling mice or cell culture lines. In addition, some labs use fluorescent antibody testing to detect viral antigen in peripheral blood smears. This procedure reportedly allows rapid and early confirmation of the disease [97]. Also, PCR technology may be used to detect CTF in blood or tissues. No specific treatment is available.

## 4.11   Tick-Borne Relapsing Fever (TBRF)

### 4.11.1   Introduction and Medical Significance

TBRF is a systemic spirochetal disease characterized by periods of fever lasting 2–9 d alternating with afebrile periods of 2–4 d. The disease is endemic across central Asia, northern Africa, tropical Africa, parts of the Middle East, and North and South America [98]. Symptoms include high fever, headache, prostration, myalgias, and sometimes gastrointestinal manifestations. Untreated, the mortality rate is between 2% and 10%. Several hundred cases are reported worldwide each year, with about 20 of those cases being diagnosed in the United States (primarily in Washington, Oregon, and northern California) [99]. Outbreaks in the western United States have most often been associated with mountain cabins or rented state or federal park cabins [100–103].

### 4.11.2   Clinical and Laboratory Findings

After an incubation period of about 8 d (range 5–15), patients with TBRF usually begin to have recurrent bouts of fever (Fig. 4.28). The total number of relapses can vary from 1 to 10 (sometimes more), lasting a week or more each time. The relapsing nature of this illness is thought to be related to various antigenic variants. As an

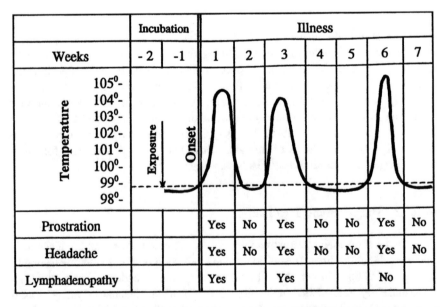

**Fig. 4.28** Recurring clinical symptoms of a 13-yr-old boy with TBRF (redrawn in part from Thompson et al. (Ref. 101))

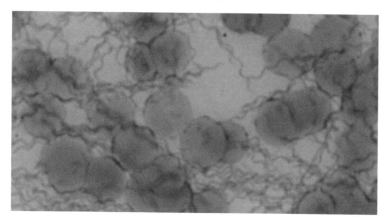

**Fig. 4.29** Relapsing fever spirochetes in blood

immune response develops to the predominant antigenic strain, variant strains multiply and cause a recrudescent infection. Transitory petechial rashes are common during the initial febrile period. Gastroenteritis-like symptoms may accompany infection. In some cases, there may be meningeal inflammation and peripheral facial palsy [104, 105]. High perinatal mortality may also result from TBRF infection; one study in Africa showed the total loss of pregnancies including abortions to be 475/1000 [106]. Laboratory findings may include neutrophilic pleocytosis of the CSF, peripheral leukocytosis, thrombocytopenia, and hypophosphatemia. Diagnosis is usually made by demonstration of the spirochetes in dark-field preparations of fresh blood or stained thick or thin blood films (Fig. 4.29) or by intraperitoneal inoculation of laboratory rats or mice with blood taken during the febrile period. When scanning fresh blood samples by dark-field microscopy, spirochetes can be readily detected under low power (400×) because of the organisms' characteristic locomotion consisting of helical rotation and twisting movements in both directions. TBRF spirochetes have an affinity for acid dyes and stain readily with aniline dyes. Giemsa stain is used most often for staining spirochetes in thick and thin film preparations. Although not widely available, serologic testing for TBRF may be aided by an ELISA. The CDC's Division of Vector-Borne Infectious Diseases, Center for Infectious Diseases, can be consulted for help with suspicious febrile illnesses.

### 4.11.3 The Etiologic Agent and Its Relationship to Louse-Borne Relapsing Fever

Relapsing fever is both tick-borne and louse-borne. Louse-borne relapsing fever is caused by *Borrelia recurrentis* and is called epidemic relapsing fever. The tick-borne disease, endemic relapsing fever, is said to be caused by many different species of *Borrelia* closely related to *B. recurrentis*. For example, *Borrelia hermsii* is the spirochete found in the tick, *Ornithodoros hermsi*; *Borrelia turicatae* is the one found in *Ornithodoros turicata*; and so forth. The idea is that each strain of *B. recurrentis* is "tick-adapted" to the point of being a distinct entity (species). Some scientists disagree, saying that all relapsing fever in humans—louse-borne and tick-borne—is caused by the same organism, *B. recurrentis* or various tick-adapted strains thereof. It becomes a matter of "splitting" or "lumping" species. This author would prefer to call them all *B. recurrentis*. However, there may be merit in retaining the various tick-borne "species" names for epidemiological labeling purposes. For example, using the name *B. hermsi* helps the reader know that we are talking about the TBRF spirochete associated with the tick, *O. hermsi*.

### 4.11.4  Ecology of TBRF

TBRF spirochetes are transmitted to humans by several species of soft ticks in the genus *Ornithodoros*. Soft ticks are not commonly encountered by people in the United States (they are not the ones that firmly attach to dogs, cats, horses, cows, humans, and so forth). They are leathery, wrinkled, or granulated organisms, often grayish in color, which live in deserts, or under dry conditions in wet climates, hiding in crevices or burrowing into loose soil. Soft ticks are adapted for feeding rapidly and leaving promptly; they are rarely ever collected on a host. They can survive many years without a blood meal. Since soft ticks generally feed for only a short period of time (30 min or so), the victim may be unaware of any recent tick bites. Rodents and other mammals serve as a natural source of infection for ticks, and transmission is by tick bite (saliva) and also sometimes through contamination of the bite wound with infective coxal fluid produced by feeding ticks just before they detach. Transstadial and transovarial transmission of the agent occurs readily. Thus, the ticks are reservoirs of infection. Geographic foci of TBRF infection are restricted to *Ornithodoros*-infested areas, such as huts, caves, log cabins, cattle barns, and uninhabited houses.

Known vectors of TBRF in the western United States include *O. hermsi*, *O. turicata*, and *O. parkeri*. *O. hermsi* is a rodent parasite that is widespread in the Rocky Mountain and Pacific Coast states of California, Nevada, Idaho, Oregon, Utah, Arizona, Washington, and Colorado, and in British Columbia, Canada (Figs. 4.30 and 4.31). They are often found infesting corners and crevices of vacation or summer cabins. *O. turicata* is found in the south central and southwestern states of Texas, New Mexico, Oklahoma, Kansas, California, Colorado, Arizona, Utah, and (spotty areas of) Florida, extending southward into Mexico (Figs. 4.32 and 4.31). There is an isolated population of them in Florida. This species is often found in burrows used by rodents or burrowing owls. In Central and South America, *Ornithodoros rudis* is considered the most important vector. It feeds on domestic birds and humans. In Africa, *Ornithodoros moubata* and *Ornithodoros erraticus* are proven vectors (Fig. 4.33). *O. moubata* transmits *B. duttonii* and feeds on humans,

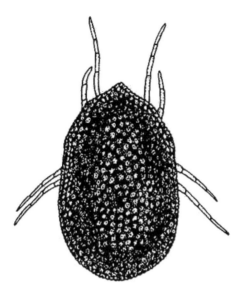

**Fig. 4.30** Adult *O. hermsi*, one of the principal vectors of relapsing fever in the western United States (US Air Force figure)

warthogs, domestic pigs, antbears, and porcupines. It is often found in cracks in walls and in earthen floors of huts.

### 4.11.5 Treatment and Control

Tetracyclines are effective against TBRF. Oral doxycycline or tetracycline for 7–14 days is a good option for treatment [46]. Non-tetracyclines include erythromycin and amoxicillin. Jarisch–Herxheimer reactions often follow treatment and should be considered [46, 107]. Prevention of relapsing fever consists of avoiding tick-infested areas or, when this is not possible, reducing the possibility of tick bites by using repellents or insecticides (see Appendix 3). Additional measures include fumigating rodent nesting sites in human habitations, "rodent-proofing" buildings in endemic areas, and eliminating rodent access to unnatural food sources.

## 4.12 Tick Paralysis

### 4.12.1 Introduction and Medical Significance

Tick paralysis is characterized by an acute, ascending, flaccid motor paralysis that may terminate fatally if the tick is not located and removed. The causative agent is believed to be a salivary toxin produced by ticks when they feed. In the strictest

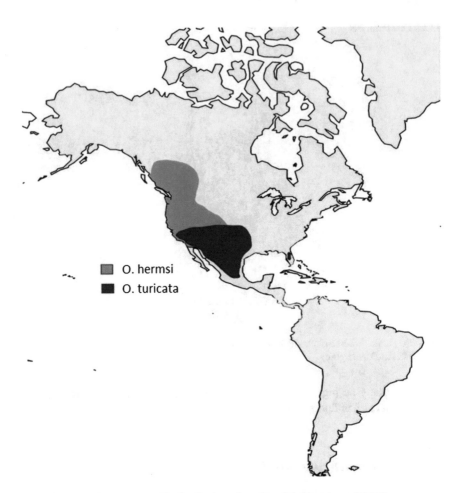

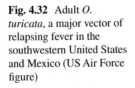

O. hermsi

O. turicata

**Fig. 4.31** Approximate geographic distribution of two New World vectors of TBRF

**Fig. 4.32** Adult *O.
turicata*, a major vector of
relapsing fever in the
southwestern United States
and Mexico (US Air Force
figure)

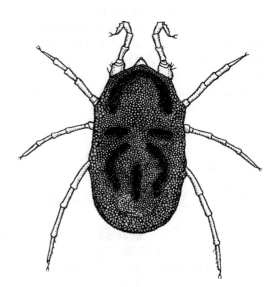

**Fig. 4.33** Approximate geographic distribution of Old World vectors of TBRF

sense, tick paralysis is not a zoonosis; however, many contend that zoonoses should include not only infections that humans acquire from animals but also diseases induced by noninfective agents, such as toxins and poisons [108]. The disease is more common than one might think (Fig. 4.34). In North America, hundreds of cases have been documented from the Montana–British Columbia region [109, 110]. It occurs in the southeastern United States as well; six cases were seen at the University of Mississippi Medical Center over a 5-yr period [111]. Clusters of tick paralysis may occur [112]. Tick paralysis is especially common in Australia. However, sporadic cases may occur in Europe, Africa, and South America.

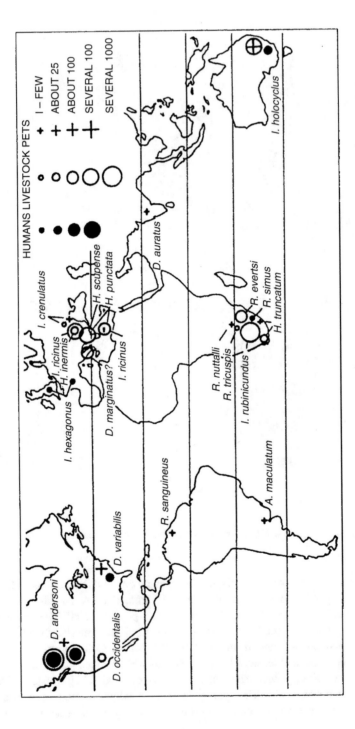

**Fig. 4.34** World distribution of tick paralysis showing approximate numbers of humans, livestock, and pets affected, and species of ticks involved. (From Gregson, JD, Canada Department of Agriculture Monograph No. 9, 1973).

## 4.12.2  Clinical Features

The site of tick bite in a case of tick paralysis looks no different from that in cases without paralysis. There is a latent period of 4–6 d before the patient becomes restless and irritable. Within 24 h, there is an acute ascending lower motor neuron paralysis of the Landry type. It usually begins with weakness of the lower limbs, progressing in a matter of hours to falling down and obvious incoordination, which is principally owing to muscle weakness, although rarely there may also be true ataxia [113]. Finally, cranial nerve weakness with dysarthria and dysphagia leads to bulbar paralysis, respiratory failure, and death. In children, presenting features may include restlessness, irritability, malaise, and sometimes anorexia and/or vomiting [113]. A tick may be found attached to the patient, usually on the head or neck. Some controversy occurs over whether or not severity of symptoms is related to the proximity of the attached tick to the patient's brain [114]. In one study, the case fatality rate in patients with ticks attached to the head or neck was higher than that in patients with ticks attached elsewhere; however, the difference was not statistically significant [110]. Although ticks causing paralysis are often attached to the head or neck, it must be noted that cases of paralysis may occur from tick bites anywhere on the body (published examples—external ear, breast, groin, and back [113]). Once the tick is found and removed, all symptoms usually disappear rapidly (but there are exceptions—see Sect. 4.12.4).

## 4.12.3  Ticks Involved and Mechanism of Paralysis

As many as 43 tick species in 10 genera have been incriminated in tick paralysis in humans, other mammals, and birds [115]. However, human cases of the malady mostly occur in only a few geographic regions, caused by three main tick species. In the northwestern United States and British Columbia region of North America, the Rocky Mountain wood tick, *D. andersoni*, is the principal tick involved (Figs. 4.6 and 4.7). This tick is an avid human biter and also is known to be a vector of RMSF organisms and CTF virus. In the southeastern United States, a kissing cousin of the Rocky Mountain wood tick, *D. variabilis*, known as the American dog tick, is the main cause of tick paralysis (Figs. 4.8 and 4.9). This tick, commonly found on dogs, cats, and other medium-sized mammals, is also a common human biter in the summer months. Human cases in Australia are primarily caused by the Australian paralysis tick, *I. holocyclus* (Figs. 4.35 and 4.36). This species is found primarily in heavily vegetated rain forest areas of eastern coastal Australia where the bandicoot is one of its main natural hosts. Beside humans and bandicoots, it also bites sheep, cattle, dogs, cats, other mammals, and birds. Another species involved in human paralysis in Australia is *Ixodes cornuatus*.

**Fig. 4.35** Adult female *I. holocyclus* tick, primary cause of tick paralysis in Australia (USAF Publ. USAFSAM-89-2)

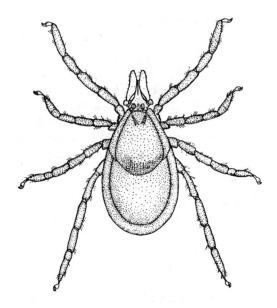

Interestingly, not all feeding female ticks—even of the species known to cause paralysis—produce paralysis. Why, out of hundreds of tick bites, does one result in paralysis? There is some evidence that in cattle, sheep, and dogs, numerous ticks feeding simultaneously (to reach a minimum dose) is necessary to elicit paralysis [109]. In humans, however, one tick is usually involved.

Most researchers believe that tick paralysis is caused by a toxin, but its nature is not well characterized. Generally, it is thought that the toxin is produced in the salivary glands of the female tick as she feeds. One alternative view would be that the toxin is produced in tick ovaries and subsequently passes to the salivary glands during later stages of tick engorgement. Although the vast majority of cases are owing to female ticks, there are reports of male ticks causing limited paralysis. This fact seems to argue against the ovary toxin theory. There are other theories for the cause of the paralysis, such as host reactions to components of the tick saliva or possibly symbiotic rickettsial organisms commonly found in tick salivary glands.

### 4.12.4   Prevention and Treatment

Since paralysis does not usually develop until late in the feeding phase of the tick (several days), frequent examination of the body and removal of any attached ticks reduce the risk of paralysis (Fig. 4.10). In the United States (*Dermacentor* ticks), after onset of paralysis, removing the tick generally results in rapid improvement—often almost miraculous. However, patients in deep paralysis should be under constant

**Fig. 4.36**  Approximate geographic distribution of *I. holocyclus*

surveillance even after tick removal, since adverse developments may still rarely occur. One report detailed a 2-yr-old who stopped breathing 32 h after tick removal [109]. The usual rapid improvement after tick removal is not always the case for *I. holocyclus* ticks in Australia—the patient may eventually die anyway. Alexander [113] says that in Australian cases, symptoms may progress for up to 2 d after tick removal before recovery sets in. As far as specific treatment goes, there is none, other than reports of some success in Australia with dog antiparalysis serum [116].

**Facts You Should Know About "Alpha-Gal," the Newly Described, Delayed Red Meat Allergy**

Recent research supports a link between bites of certain ticks and allergy to red meat. The ticks responsible for this condition are lone star ticks (*Amblyomma americanum*) (Fig. 4.17), which are common in the southern US. In sensitized individuals, allergy to a substance "alpha-gal" found in red meat triggers allergic reactions several hours following red meat consumption.

1. Alpha-gal is short for galactose-α-1,3-galactose. Alpha-gal is a carbohydrate expressed on the surface of glycolipids and glycoproteins of most mammals. Humans lack *3-galactosyltransferase* required for alpha-gal synthesis, so they cannot manufacture or express alpha-gal themselves. However, they are still capable of producing IgG, IgM, and IgE antibodies to alpha-gal when exposed to it.

2. Tick bites appear to cause alpha-gal sensitivity and can trigger patients to make IgE to alpha-gal. This results in IgE sensitization to alpha-gal. Although the exact mechanism is unknown, isotype switching to IgE is thought to occur in the skin following a bite from either larval or adult ticks. All are potential sources of alpha-gal that can trigger production of IgE and sensitization to it. Additional supporting evidence that tick bites are the major source for production of alpha-gal-specific IgE include IgE to alpha-gal increasing in humans after tick bites, reports of pruritus after tick bites correlating with the presence of alpha-gal in human serum, and IgE to alpha-gal correlating with the presence of IgE in lone star tick extract. In areas where lone star ticks are not present, IgE to alpha-gal is absent.

3. The lone star tick and resulting "alpha-gal" allergy are found in the southeastern United States. The lone star tick, which serves as the cause or "vector" for alpha-gal sensitization, is found primarily in the southeastern United States, but over the years, this geographic distribution has expanded as far north as Maine.

4. Alpha-gal is present on the surface of glycolipids and glycoproteins (including chylomicrons), which are slowly absorbed into human circulation; therefore, the delay in symptoms after meat exposure in allergic individuals is thought to be related to slow mediator release triggered by low-density lipoproteins (LDL) and very low-density lipoproteins (VLDL), the metabolic breakdown of products of these chylomicrons. Alpha-gal is present on the surface of LDL and VLDL. In sensitized individuals (who exhibit IgE antibodies to alpha-gal), cross-linking IgE results in mast cell degranulation and clinical symptoms. Because chylomicron packaging and lipid absorption into the lymphatic system takes time, clinical manifestations of the resulting alpha-gal allergic reaction do not appear for several hours.

(continued)

5. History is key to suspecting delayed red meat allergy. Delayed urticarial, angioedema, or anaphylaxis after consuming red meat should be prompt consideration for alpha-gal allergy as part of the differential diagnosis. Unlike classic food allergies which typically present 5–30 min after ingestion, patients allergic to red meat because of alpha-gal allergy generally do not become symptomatic until 3–6 h following red meat ingestions. Patients may have reactions that awaken them from sleep. Because the time between symptom onset and ingestion of alpha-gal may be separated by hours, the association can be difficult and may be overlooked. The clinical picture can be further complicated since they may tolerate red meat for years before the symptoms begin and every red meat ingestion might not trigger a reaction. Any time a delayed red meat allergy is suspected, questions about a history of previous tick bites are appropriate. *Amblyomma americanum* bites can cause intense pruritus that can continue for weeks, which makes them memorable for many patients.

6. Many animal meats that express alpha-gal fall into the "red meat" category. These animals include beef, pork, lamb, squirrel, horse, goat, venison, kangaroo, seal, and whale. All of these should be avoided in patients with alpha-gal allergy. Patients sensitized to alpha-gal should tolerate chicken, turkey, and fish. Venison prepared by a hunter contains minimal fat and might be tolerated versus venison preparations by a butcher that might include pork fat and cause a reaction.

7. If the history suggests alpha-gal allergy, testing is available but results must be interpreted with appropriate clinical correlation. The initial diagnostic approach includes obtaining both IgE levels to alpha-gal and to the suspected individual red meats containing alpha-gal. Serum IgE to alpha-gal should be present, but changes in titers are not directly correlated to symptom timing nor severity. Also, patients who have not experienced red meat allergy symptoms may still test positive for IgE titers to alpha-gal and beef. This creates the necessity for clinical correlation to prevent misdiagnosis of red meat allergy. The typical patient with alpha-gal allergy has elevated IgE titers to galactose-$\alpha$-1,3-galactose, beef, lam, and pork with negative IgE titers to chicken, turkey, and fish. Despite oral food challenge being the gold standard for classic food allergy, it is not currently recommended for alpha-gal allergy diagnosis due to the delayed nature of the reactions.

8. There is cross-reactivity between alpha-gal and other proteins including those in cow's milk and gelatin. Alpha-gal is a cross-reactive carbohydrate determinant that may cause patients to demonstrate specific IgE to cat, dog and cow's milk proteins; this cross-reactivity may not be clinically significant. On the other hand, gelatin from mammals is found in multiple foods and could serve as a covert trigger for allergic symptoms.

(continued)

Gelatin-containing foods include confectioneries (marshmallows, food thickeners, glazes, icing) and fat substitutes (yogurt, mayonnaise, ice cream, sausage coatings, salami, tinned hams, meat stock). Gelatin can also be found in certain vaccines such as influenza, MMR, and varicella, but alpha-gal sensitization relative to tolerance of gelatin-containing vaccines has not been investigated. It is important to keep these cross-reactive foods in mind if a patient has been identified as red meat allergic and has eliminated red meats but remains symptomatic.

9. Treatment recommendations for alpha-gal allergy include avoiding consumption of all red meats and carrying an epinephrine auto-injector. As in classic food allergy, cooking methods (rare vs well done) would not be expected to alter red meat tolerance. All red meats previously listed should be avoided, and if accidentally ingested, reactions should be treated accordingly. Anaphylaxis should be treated with intramuscular epinephrine.

Because the natural course of alpha-gal allergy is currently unknown, the decrease in alpha-gal titers seen over time has not been correlated to future reactivity.

10. Tick bite avoidance is the only known method to prevent alpha-gal sensitization. Suggestion for avoiding tick bites include:

- Wear protective clothing, preferably treated with 0.5% permethrin.
- Use repellents with 20–30% N,N-diethyl-meta-toluamide (DEET) on exposed skin surfaces.
- Avoid tall grass/leaf litter and travel along the center of trails.
- Carefully examine gear and pets and remove any ticks upon returning indoors.
- Tumble clothes in a dryer for 1 hour using high heat to kill any remaining ticks.
- Use extra precautions during March through September when ticks are most active.

Adapted from the article, "Ramey K, Stewart PH. Top ten facts you should know about "alpha-gal," the newly described delayed red meat allergy. J Mississippi St Med Assoc. 2016;57(9):279–81. Used with permission.

# References

1. Raoult D, Olson JG. Emerging rickettsioses. In: Scheld WM, Craig WA, Hughes JM, editors. Emerging infections, vol. 3. Washington, DC: ASM Press; 1999. p. 17–35.
2. Goddard J, Norment BR. Spotted fever group rickettsiae in the lone star tick. J Med Entomol. 1986;23:465–72.
3. Mixon TR, Campbell SR, Gill JS, Ginsberg HS, Reichard MV, Schulze TL, et al. Prevalence of *Ehrlichia, Borrelia,* and *Rickettsial* agents in *Amblyomma americanum* collected from nine states. J Med Entomol. 2006;43:1261–8.

4. Schriefer ME, Azad AF. Changing ecology of Rocky Mountain spotted fever. In: Sonenshine DE, Mather TN, editors. Ecological dynamics of tickborne zoonoses. New York: Oxford University Press; 1994. p. 314–24.

5. Harden VA. Rocky Mountain spotted fever: history of a twentieth-century disease. Baltimore: Johns Hopkins Press; 1990.

6. Ricketts H. The transmission of Rocky Mountain spotted fever by the bite of the wood tick. J Am Med Asoc. 1906;47:358.

7. Spencer RR, Parker RR. Rocky Mountain spotted fever: vaccination of monkeys and man. Pub Health Rep. 1925;40:2159–67.

8. Burgdorfer W, Barbour AG, Hayes SF, Benach JL, Grunwaldt E, Davis JP. Lyme disease- a tick-borne spirochetosis? Science. 1982;216:1317–9.

9. CDC. Summary of notifiable infectious diseases and conditions – United States, 2015. CDC, MMWR. 2017;64(53):1–144.

10. Heymann DL, editor. Control of communicable diseases manual. 20th ed. Washington, DC: American Public Health Association; 2015.

11. Kirk JL, Fine DP, Sexton DJ, Muchmore HG. Rocky Mountain spotted fever: a clinical review based on 48 confirmed cases, 1943–1986. Medicine. 1990;69:35–45.

12. CDC. Diagnosis and management of tickborne rickettsial diseases: Rocky Mounatin spotted fever, ehrlichioses, and anaplasmosis. MMWR. 2006;55(RR-4):1–29.

13. Spach DH, Liles WC, Campbell GL, Quick RE, Anderson DEJ, Fritsche TR. Tick-borne diseases in the United States. N Engl J Med. 1993;329:936–47.

14. Demma LJ, Traeger MS, Nicholson WL, Paddock CD, Blau DM, Eremeeva ME, et al. Rocky Mountain spotted fever from an unexpected tick vector in Arizona. N Engl J Med. 2005;353(6):587–94.

15. Goddard J, Layton MB. A Guide to Ticks of Mississippi. Mississippi Agriculture and Forestry Experiment Station, Mississippi State University, Bulletin Number 1150; 2006. 17 pp.

16. James AM, Freier JE, HKeirans JE, Durden LA, Mertins JW, Schlater JL. Distribution, seasonality, and hosts of the Rocky Mountain wood tick in the United States. J Med Entomol. 2006;43:17–24.

17. Anonymous. Doxycycline for young children? Med Lett Drugs Ther. 2016;58:75.

18. Todd SR, Dahlgren FS, Traeger MS, Beltran-Aguilar ED, Marianos DW, Hamilton C, et al. No visible dental staining in children treated with doxycycline for suspected Rocky Mountain Spotted Fever. J Pediatr. 2015;166(5):1246–51.

19. CDC. Consequences of delayed diagnosis of Rocky Mountain spotted fever in children – West Virginia, Michigan, Tennessee, and Oklahoma, May through July, 2000. CDC, MMWR. 2000;49:885–6.

20. Lackman DB, Parker RR, Gerloff RK. Serological characteristics of a pathogenic rickettsia occurring in *Amblyomma maculatum*. Public Health Rep. 1949;64:1342–9.

21. Parker RR. A pathogenic rickettsia from the Gulf Coast tick, *Amblyomma maculatum*. Proceedings of the third international congress on microbiology, New York;, 1940. p. 390–1.

22. Parker RR, Kohls GM, Cox GW, Davis GE. Observations on an infectious agent from *Amblyomma maculatum*. Public Health Rep. 1939;54:1482–4.

23. Goddard J. Experimental infection of lone star ticks, *Amblyomma americanum* (L.), with *Rickettsia parkeri* and exposure of Guinea pigs to the agent. J Med Entomol. 2003;40:686–9.

24. Paddock CD, Sumner JW, Comer JA, Zaki SR, Goldsmith CS, Goddard J, et al. *Rickettsia parkeri* – a newly recognized cause of spotted fever rickettsiosis in the United States. Clin Infect Dis. 2004;38:805–11.

25. Ekenna O, Paddock CD, Goddard J. Gulf Coast tick rash illness caused by *Rickettsia parkeri*. J Mississippi St Med Assoc. 2014;55:216–9.

26. Finley RW, Goddard J, Raoult D, Eremeeva ME, Cox RD, Paddock CD. *Rickettsia parkeri*: a case of tick-borne, eschar-associated spotted fever in Mississippi. International conference on emerging infectious diseases, Atlanta, GA, 2006 March 19–22, Abstract No 188; 2006.

27. Whitman TJ, Richards AL, Paddock CD, Tamminga CL, Sniezek PJ, Jiang J, et al. *Rickettsia parkeri* infection after tick bite, Virginia. Emerg Infect Dis. 2007;13:334–5.

28. Paddock CD, Goddard J. The evolving medical and veterinary importance of the Gulf Coast tick. J Med Entomol. 2015;52:230–52.
29. Goddard J. American Boutonneuse Fever – a new spotted fever rickettsiosis. Inf Med. 2004;21:207–10.
30. Paddock CD, Allerdice ME, Karpathy SE, Nicholson WL, Levin ML, Smith TC, et al. Isolation and characterization of a unique strain of *Rickettsia parkeri* associated with the hard tick *Dermacentor parumapertus* Neumann in the western United States. Appl Environ Microbiol. 2017;17(10). https://doi.org/10.1128/AEM.03463-16.
31. Goddard J. A ten-year study of tick biting in Mississippi: implications for human disease transmission. J Agromedicine. 2002;8:25–32.
32. Moraru GM, Goddard J, Murphy A, Link D, Belant JL, Varela-Stokes A. Evidence of antibodies to spotted fever group rickettsiae in small mammals and quail from Mississippi. Vector Borne Zoonotic Dis. 2013;13(1):1–5.
33. Bacellar F, Beati L, Franca A. Israeli spotted fever rickettsia associated with human disease in Portugal. Emerg Inf Dis. 1999;5:835–6.
34. Raoult D, Fournier P, Abboud P, Caron F. First documented human *Rickettsia aeschlimannii* infection. Emerg Infect Dis. 2002;8:748–9.
35. Raoult D. Rickettsioses as paradigms of new or emerging infectious diseases. Clin Microbiol Rev. 1997;10:694–719.
36. Walker DH, Dumler JS. Emergence of the ehrlichioses as human health problems. Emerg Infect Dis. 1996;2:18–28.
37. Maeda K, Markowitz N, Hawley RC, Ristic M, Cox D, McDade JE. Human infection with *Ehrlichia canis* a leukocytic rickettsia. N Engl J Med. 1987;316:853–6.
38. Dumler JS, Bakken JS. Ehrlichial diseases of humans: emerging tick-borne infections. Clin Infect Dis. 1995;20:1102–10.
39. Harris RM, Couturier BA, Sample SC, Coulter KS, Casey KK, Schlaberg R. Expanded geographic distribution and clinical characteristics of *Ehrlichia ewingii* infections, United States. Emerg Infect Dis. 2016;22(5):862.
40. Pritt BS, Sloan LM, Johnson DKH, Munderloh UG, Paskewitz SM, McElroy KM, et al. Emergence of a new pathogenic Ehrlichia species, Wisconsin and Minnesota, 2009. New Engl J Med. 2011;365(5):422–9.
41. Wormser GP, Pritt B. Update and commentary on four emerging tick-borne infections: *Ehrlichia muris*-like agent, *Borrelia miyamotoi*, deer tick virus, Heartland virus, and whether ticks play a role in transmission of *Bartonella henselae*. Infect Dis Clin North Am. 2015;29(2):371–81.
42. Dumler JS. Ehrlichiosis and anaplasmosis. In: Guerrant RL, Walker DH, Weller PF, editors. Tropical infectious diseases: principles, pathogens, and practice. 3rd ed. New York: Saunders Elsevier; 2011. p. 339–44.
43. CDC. Anaplasmosis and ehrlichiosis – Maine, 2008. CDC, MMWR. 2009;58(37):1033–6.
44. Biggs H, Behravesh CB, Bradley KK, Dahlgren FS, Drexler NA, Dumler JS, et al. Diagnosis and management of tickborne rickettsial diseases: Rocky Mountain spotted fever and other spotted fever group rickettsioses, ehrlichioses, and anaplasmosis – United States. CDC, MMWR, R&R. 2016;65:1–45.
45. Holman MS, Caporale DA, Goldberg J, Lacombe E, Lubelczyk C, Rand PW, et al. *Anaplasma phagocytophilum, Babesia microti,* and *Borrelia burgdorferi* in *Ixodes scapularis* in southern coastal Maine. Emerg Infect Dis. 2004;10:744–6.
46. Bope ET, Kellerman R. Conn's current therapy. Philadelphia: Elsevier Saunders; 2017. 1375 p
47. Pritt BS, Respicio-Kingry LB, Sloan LM, Schriefer ME, Replogle AJ, Bjork J, et al. *Borrelia mayonii* sp. nov., a member of the *Borrelia burgdorferi sensu lato* complex, detected in patients and ticks in the upper midwestern United States. Int J Syst Evol Microbiol. 2016;66(11):4878–80.
48. Ginsberg HS, Faulde MK. Ticks. In: Bonnefoy X, Kampen H, Sweeney K, editors. Public health significance of urban pests. Copenhagen: WHO Regional Office for Europe; 2008. p. 303–45.

49. Kuehn BM. CDC estimates 300,000 U.S. cases of Lyme disease annually. JAMA. 2013;310:1110.
50. Surveillance CDC. For Lyme disease – United States, 2008–2015. CDC, MMWR, Surveillance Summaries. 2017;66:1–13.
51. Steere AC, Malawista SE, Snydman DR, Shope RE, Andiman WA, Ross MR, et al. Lyme arthritis: an epidemic of oligoarticular arthritis in children and adults in three Connecticut communities. Arth Rheum. 1977;20:7–17.
52. Steere AC, Grodzicki RL, Kornblatt AN, Craft JE, Barbour AG, Burgdorfer W, et al. The spirochetal etiology of Lyme disease. N Engl J Med. 1983;308:733–40.
53. Spielman A, Clifford CM, Piesman J, Corwin MD. Human babesiosis on Nantucket island, USA: description of the vector, *Ixodes dammini* N.SP. J Med Entomol. 1979;15:218–34.
54. Oliver JH, Owsley MR, Hutcheson HJ, James AM, Chunsheng C, Irby WS, et al. Conspecificity of the ticks *Ixodes scapularis* and *Ixodes dammini*. J Med Entomol. 1993;30:54–63.
55. Keirans JE, Hutcheson HJ, Durden LA, Klompen JSH. *Ixodes scapularis*: redescription of all active stages, distribution, hosts, geographical distribution, and medical and veterinary importance. J Med Entomol. 1996;33:297–318.
56. Goddard J. Not all erythema migrans lesions are Lyme disease. Am J Med. 2016.;epub ahead of print. https://doi.org/10.1016/j.amjmed:doi:10.1016/j.amjmed.
57. Nadelman RB. Tick-borne diseases: a focus on Lyme disease. Inf Med. 2006;23:267–80.
58. Rahn DW. Lyme disease – where's the bug? N Engl J Med. 1994;330:282–3.
59. Magnarelli LA. Current status of laboratory diagnosis for Lyme disease. Am J Med. 1995;98(suppl 4A):10s–4s.
60. Goddard J, Piesman J. New records of immature *Ixodes scapularis* from Mississippi. J Vector Ecol. 2006;31(2):421.
61. Masters EJ, Donnell HD, Fobbs M. Missouri Lyme disease: 1989 through 1992. J Spiro Tick-Borne Dis. 1994;1:12–3.
62. Goddard J, Varela-Stokes A, Finley RW. Lyme-disease-like illnesses in the South. J Mississippi State Med Assoc. 2012;53(3):68–72.
63. Farlow J, Wagner DM, Dukerich M, Stanley M, Chu M, Kubota K, et al. *Francisella tularensis* in the United States. Emerg Infect Dis. 2005;11:1835–41.
64. Olano JP, Peters CJ, Walker DH. Distinguishing tropical infectious diseases from bioterrorism. In: Guerrant RL, Walker DH, Weller PF, editors. Tropical infectious diseases: principles, pathogens, and practice, vol. 2. Philadelphia: Churchill Livingstone; 2006. p. 1380–99.
65. Markowitz LE, Hynes NA, de la Cruz P, Campos E, Barbaree JM, Plikaytis BD, et al. Tick-borne tularemia. J Am Med Assoc. 1985;254:2922–5.
66. Haake DA. Tularemia. In: Rakel RE, editor. Conn's current therapy. Philadelphia: W.B. Saunders; 1997. p. 166–8.
67. Francis E, Mayne B. The occurrence of tularemia in nature as a disease of man. Public Health Rep. 1921;36:1731–8.
68. Hopla CE. The ecology of tularemia. Adv Vet Sci Comp Med. 1974;18:25–53.
69. Parker RR, Spencer RR, Francis E. Tularemia infection in ticks of the species *Dermacentor andersoni* in the Bitteroot Valley, Montana. Public Health Rep. 1924;39:1052–73.
70. Hopla CE. The transmission of tularemia organisms by ticks in the southern states. South Med J. 1960;53:92–7.
71. Cross JT. Tularemia in the United States. Inf Med. 1997;14:881–90.
72. Enderlin G, Morales L, Jacobs RF, Cross JT. Streptomycin and alternative agents for the treatment of tularemia: review of the literature. Clin Infect Dis. 1994;19:42–7.
73. Homer M, Agular-Delfin I, Telford SRI, Krause PJ, Persing DH. Babesiosis. Clin Microbiol Rev. 2000;13:451–69.
74. Telford SRI, Weller PF, Maquire JH. Babesiosis. In: Guerrant RL, Walker DH, Weller PF, editors. Tropical infectious diseases: principles, pathogens, and practice. 3rd ed. New York: Saunders Elsevier; 2011. p. 676–81.
75. Gorenflot A, Moubri K, Precigout E, Carcy B, Schetters TP. Human babesiosis. Ann Trop Med Parasitol. 1998;92:489–501.

76. CDC. Babesiosis – Connecticut. CDC, MMWR. 1989;38:649–50.
77. Markell E, Voge M, John D. Medical parasitology. 7th ed. Philadelphia: W.B. Saunders; 1992.
78. Spielman A, Wilson ML, Levine JF, Piesman JF. Ecology of *Ixodes dammini*-borne human babesiosis and Lyme disease. Annu Rev Entomol. 1985;30:439–60.
79. Clark IA, Jacobson LS. Do babesiosis and malaria share a common disease process. Ann Trop Med Parasitol. 1998;92:483–8.
80. Thomford JW, Conrad PA, Telford SR III, Mathiesen D, Eberhard ML, Herwaldt BL, et al. Cultivation and phylogenetic characterization of a newly recognized human pathogenic protozoan. J Infect Dis. 1994;169:1050–6.
81. Hedayti T, Martin R. Babesiosis. e-medicince. http://www.emedicine.com/EMERG/topic49. htm; 2007.
82. Ternovoi VA, Protopopova EV, Chausov EV, Novikov DV, Leonova GN, Netesov SV, et al. Novel variant of tickborne encephalitis, Russia. Emerg Infect Dis. 2007;13:1574–8.
83. McMullan LK, Folk SM, Kelly AJ, MacNeil A, Goldsmith CS, Metcalfe MG, et al. A new Phlebovirus associated with severe febrile illness in Missouri. N Engl J Med. 2012;367(9):834–41.
84. Pastula DM, Turabelidze G, Yates KF, Jones TF, Lambert AJ, Panella AJ, et al. Heartland virus disease – United States, 2012–2013. CDC. MMWR. 2014;63:270–1.
85. Kosoy OI, Lambert AJ, Hawkinson DJ, Pastula DM, Goldsmith CS, Hunt DC, et al. Novel thogotovirus associated with febrile illness and death, United States, 2014. Emerg Infect Dis. 2015;21(5):760–4.
86. Savage HM, Burkhalter KL, Godsey MSJ, Panella NA, Ashley DC, Nicholson WL, et al. Bourbon virus in field-collected ticks, Missouri, USA. Emerg Infect Dis. 2017;23(12):2017–22.
87. Monath TP, Johnson KM. Diseases transmitted primarily by arthropod vectors. In: Last JM, Wallace RB, editors. Public Health and Preventive Medicine. 13th ed. Norwalk: Appleton and Lange; 1992.
88. Gresikova M, Calisher CH. Tick-borne encephalitis. In: Monath TP, editor. The arboviruses: epidemiology and ecology, vol. 4. Boca Raton: CRC Press; 1989. p. 177–84.
89. Goddard J. Ticks and tick-borne diseases affecting military personnel. USAF, School of Aerospace Medicine: San Antonio; 1989. 140 p
90. Gritsun TS, Lashkevich VA, Gould EA. Tick-borne encephalitis. Antivir Res. 2003;57:129–46.
91. Hinten SR, Beckett GA, Gensheimer KF, Pritchard E, Courtney TM, Sears SD, et al. Increased recognition of Powassan encephalitis in the United States, 1999–2005. Vector-Borne Zoon Dis. 2008;8(6):733–40.
92. Nuttall PA, Labuda M. Tick-borne encephalitis subgroup. In: Sonenshine DE, Mather TN, editors. Ecological dynamics of tick-borne zoonoses. New York: Oxford University Press; 1994. p. 351.
93. Telford SR III, Armstrong PM, Katavolos P, Foppa I, Garcia ASO, Wilson ML, et al. A new tick-borne encephalitis-like virus infecting New England deer ticks, *Ixodes dammini*. Emerg Infect Dis. 1997;3:165–70.
94. Tavakoli NP, Wang H, Dupuis M, Hull R, Ebel GD, Gilmore EJ, et al. Fatal case of deer tick virus encephalitis. N Engl J Med. 2009;360(20):2099–107.
95. Aberle JH, Aberle SW, Kofler RM, Mandl CW. Humoral and cellular immune response to RNA immunization with flavivirus replicons derived from tick-borne encephalitis. J Virol. 2005;79:15107–13.
96. WHO. Requirements for tick-borne encephalitis vaccine (inactivated). World Health Organization, Geneva, Technical Report Series, No. 889; 1999. p. 44–62.
97. Emmons R. Ecology of Colorado tick fever. Ann Rev Microbiol. 1988;42:49–64.
98. Varma MGR. Ticks and mites. In: Lane RP, Crosskey RW, editors. Medical insects and arachnids. London: Chapman and Hall; 1993. p. chap. 18.
99. CDC. Tickborne relapsing fever – United States, 1990–2011. CDC, MMWR. 2015;64:58–60.
100. CDC. Outbreak of relapsing fever – Grand Canyon National Park, Arizona. CDC, MMWR. 1991;40:296–7.

101. Thompson RS, Burgdorfer W, Russell R, Francis BJ. Outbreak of tick-borne relapsing fever in Spokane County, Washington. J Am Med Assoc. 1969;210:1045–9.
102. Trevejo RT, Schriefer ME, Gage KL, Safranek TJ, Orloski KA, Pape WJ, et al. An interstate outbreak of tick-borne relapsing fever among vacationers at a Rocky Mountain cabin. Am J Trop Med Hyg. 1998;58:743–7.
103. CDC. Tickborne relapsing fever outbreak at an outdoor education Camp – Arizona, 2014. CDC, MMWR. 2015;64:651–2.
104. Cadavid D, Barbour AG. Neuroborreliosis during relapsing fever: review of the clinical manifestations, pathology, and treatment of infections in humans and experimental animals. Clin Infect Dis. 1998;26:151–64.
105. CDC. Common source outbreak of relapsing fever – California. CDC, MMWR. 1990;39:579.
106. Jongen VH, van Roosmalen J, Tiems J, Van Holten J, Wetsteyn JC. Tick-borne relapsing fever and pregnancy outcome in rural Tanzania. Acta Obstet Gynecol Scand. 1997;76:834–8.
107. Edlow JA. Tick-borne diseases, relapsing fever. e-medicine, http://www.emedicine.com/EMERG/topic590.htm; 2007.
108. Kocan AA. Tick paralysis. J Am Vet Med Assoc. 1988;192:1498–500.
109. Gregson JD. Tick paralysis: an appraisal of natural and experimental data. Canada Dept. Agri. Monograph No. 9; 1973. p. 48.
110. Schmitt N, Bowmer EJ, Gregson JD. Tick paralysis in British Columbia. Canadian Med Assoc J. 1969;100:417–21.
111. Vedanarayanan VV, Evans OB, Subramony SH. Tick paralysis in children: electrophysiology and possibility of misdiagnosis. Neurology. 2002;59:1088–90.
112. CDC. Cluster of tick paralysis cases – Colorado, 2006. CDC, MMWR. 2006;55:934–5.
113. Alexander JO. Arthropods and human skin. Berlin: Springer; 1984. 422 p
114. Stanbury JB, Huyck JH. Tick paralysis: a critical review. Medicine. 1945;24:219–42.
115. Gothe R, Kunze K, Hoogstraal H. The mechanisms of pathogenicity in the tick paralysis. J Med Entomol. 1979;16:357–69.
116. Kaire GH. Isolation of tick paralysis toxin from Ixodes holocyclus. Toxicon. 1966;4:91–7.

# Chapter 5
# Flea-Borne Diseases

## 5.1 Basic Flea Biology

Fleas have complete metamorphosis with egg, larva, pupa, and adult stages (Fig. 5.1). The adults have piercing-sucking mouthparts and feed exclusively on blood (Fig. 5.2). Hosts of fleas are domesticated and wild animals, especially wild rodents, but many species will bite people (Fig. 5.3). If hosts are available, fleas may feed several times daily, but in the absence of hosts, adults may fast for months, especially at low-to-moderate temperatures. Some species have specialized life cycles, but in general, the life cycle of most fleas ranges from 30 to 75 d.

Since cat fleas are a notable pest and seemingly ubiquitous, their life cycle is presented here (Fig. 5.1). Adult female fleas begin laying eggs 1–4 d after starting periodic blood feeding. Blood meals are commonly obtained from cats, dogs, and people, but other medium-sized mammals, such as raccoons and opossums, may be utilized as well. Females lay 10–20 eggs daily and may produce several hundred eggs in their lifetime. Eggs are normally deposited in nest litter, bedding, carpets, and so forth. Warm, moist conditions are optimal for egg production. Eggs quickly hatch into spiny, yellowish-white larvae. Flea larvae have chewing mouthparts and feed on host-associated debris, including food particles, dead skin, and feathers. Blood defecated by adult fleas also serves as an important source of nutrition for the larvae. Larvae pass through three molts (instars) prior to pupating. Flea larvae are very sensitive to moisture and will quickly die if continuously exposed to <60–70% relative humidity. Pupating flea larvae spin a loose silken cocoon interwoven with debris. If environmental conditions are unfavorable, or if hosts are not available, developing adult fleas may remain inactive within the cocoon for extended periods. Adult emergence from the cocoon may be triggered by vibrations resulting from host animal movements.

© Springer International Publishing AG, part of Springer Nature 2018
J. Goddard, *Infectious Diseases and Arthropods*, Infectious Disease,
https://doi.org/10.1007/978-3-319-75874-9_5

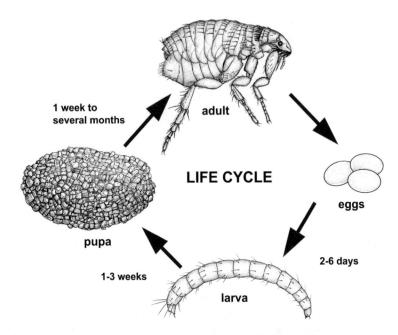

**Fig. 5.1** Flea life cycle. (Figure courtesy Mississippi State University Extension Service, with permission)

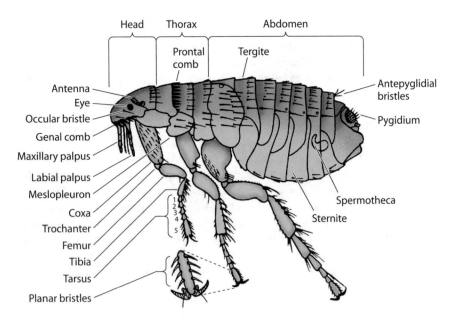

**Fig. 5.2** Diagrammatic flea with structures labeled (Centers for Disease Control figure)

**Fig. 5.3** Flea bites on the knee (Photo copyright 2001 by Jerome Goddard, Ph.D.)

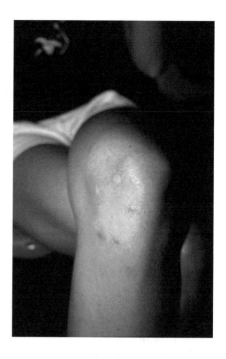

## 5.2 Plague

### 5.2.1 Introduction and Clinical Presentation

Plague, a zoonotic disease caused by the bacterium, *Yersinia pestis*, has been associated with humans since recorded history. It is a flea-transmitted disease with hundreds of cases occurring annually over much of the world (Fig. 5.4), but especially in Madagascar and countries in Africa. In the United States, ~10 cases occur each year, mostly from Arizona, California, Colorado, and New Mexico [1], although in 2015 there were 16 reported plague cases, the highest since 2006 [2]. Cases may be urban (human epidemics associated with domestic rats) or sylvatic (wild rodent populations). Sylvatic plague, sometimes also called campestral plague, is ever-present in endemic areas, circulating among rock and ground squirrels, deer mice, voles, chipmunks, and others. Transmission from wild rodents to humans is rare. *Y. pestis* inflicts damage on the host animal by an endotoxin present on its surface. Metabolism of many cell types is hampered, tissues undergo degenerative and necrotic changes, and internal hemorrhages may occur. Neural tissues and heart muscle may be damaged as well.

There are three principal forms of human plague: *bubonic*, infection of the lymph nodes; *septicemic*, infection of the blood; and *pneumonic*, infection of the lungs. After a 2- to 8-d incubation period, the disease is characterized by fever and chills, quickly followed by prostration. There is headache, the eyes are injected, and the

**Reported\* Plague Cases by Country, 2010-2015**

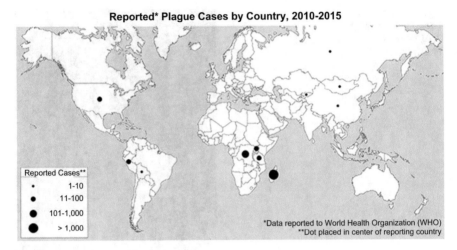

**Fig. 5.4** Distribution of plague worldwide (CDC figure)

facies are characteristic of extreme illness. Delirium appears early. The characteristic lesion, a bubo, is an extremely tender, swollen, firm, nonfluctuant lymph node in the region draining the site of the flea bite [3]. The skin overlying the node is usually erythematous, shiny, and edematous. Some patients will present with an acute febrile illness without the bubo, which is called septicemic plague. Septicemic plague may cause signs and symptoms similar to those of gastrointestinal infection, urinary tract infection, respiratory tract infection, appendicitis, or a nonspecific viral syndrome [4]. Spread of the infection to the lungs may result in pneumonic plague, which is especially dangerous and leads to human-to-human transmission by infective airborne droplets. Bubonic plague is the most common form of plague. From 1971 to 1995, it comprised 84% of 320 cases [5]. The remaining percentages were 14% septicemic (with no lymph node involvement) and ~3% primary pneumonic, acquired by inhalation of infectious aerosols from another person or animal with plague pneumonia.

## 5.2.2   History

No other disease can compare to the devastating effects of plague on human civilization. There have been at least three major pandemics of plague [6, 7]. The first, the Plague of Justinian, occurred in the sixth century and killed an estimated 40 million people. The second, called the Black Death, occurred during the fourteenth century, claiming the lives of 25 million people. The third (or modern) pandemic began in the late nineteenth century and killed an estimated 12 million people. Many other smaller epidemics have occurred, such as the London epidemic of 1666, which killed 70,000 people. The spread of plague around the world is thought to be closely related to commerce and especially rat-infested ships. During the last pandemic (nineteenth century), the outbreak started in northern China and soon reached Hong Kong via routes of commerce. It was subsequently transferred to other continents by way of rats on steamships. Northern China has been regarded as the cradle of plague, with permanent foci in wild rodents, and the source of transfer to humans and domestic (also called commensal) rodents along ancient land trade routes and hostelries to European urban centers, resulting in severe pandemics. Shipping routes have spread plague to seaports worldwide, where native rodents have become infected, starting up the cycle in those areas. It is believed that plague was introduced into the United States in the San Francisco area in 1899.

Of course, centuries ago, no one knew for sure what caused outbreaks of plague. Some thought it was contaminated soil or air; others considered plague a direct judgment of God. Fear of the disease and sometimes sheer panic altered human behavior in affected areas. Mullett [8] describes this well:

> The Black Death itself everywhere produced the most diverse effects. Its appalling mortality encouraged dissipation and asceticism, persecution and indifference. Wars were thrown off, trade and agriculture disrupted, and government suspended. Love, trust, and faithfulness took flight, and the patient was forsaken by all except his dog. Neither his nearest and dearest nor his priest and physician dared visit him. Diabolism flourished as persons paid homage to the devil, and sorcerers abounded. Flagellation, choremania, and children's pilgrimages conspicuously reflected current neuroses. Jews, as might be expected, were brutally massacred when charges of ritual murder and the deliberate distribution of a plague poison gained wholesale credence.

## 5.2.3   Ecology of Plague

Plague is maintained in the western United States in a sylvatic cycle involving resistant rodent hosts, such as ground squirrels, deer mice, and the California vole. There has been an increase in cases in recent years and an eastward shift geographically [9] (Fig. 5.5). Transmission of plague from rodent to rodent is by flea bite; several host-specific flea species may be involved. The disease becomes amplified when it spills over into susceptible species, such as prairie dogs and rock squirrels, resulting in widespread epizootics. People become ill when these susceptible hosts die (sometimes in huge die-offs involving hundreds of rodents) and their fleas subsequently bite nearby humans. Flea vectors include the Oriental rat flea, *Xenopsylla cheopis* (Fig. 5.6); the ground squirrel flea, *Oropsylla montana* (Fig. 5.7); and others [10]. In some cases, plague spreads to urban areas and involves commensal (domestic) rodents, particularly *Rattus rattus*, and their fleas such as *Xenopsylla cheopis*. However, there are other means of acquiring plague. During 1970–1996, 16% of cases were acquired by direct contact with blood or tissue of an infected animal (such as skinning a rabbit) [5]. Fairly recently, there has been an increase in the number of human cases associated with domestic cats [6]. In fact, before 1977, domestic cats were never reported as sources of human plague infection; however, since 1977, cats have been identified as the source of infection for 15 human plague cases [1]. Unlike dogs, which do not usually show signs of plague infection, cats

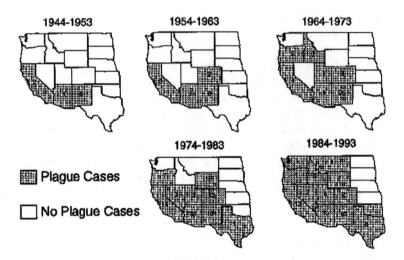

| 1944-1953 | 1954-1963 | 1964-1973 |

| 1974-1983 | 1984-1993 |

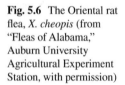

■ Plague Cases

☐ No Plague Cases

**Fig. 5.5** Number of human plague cases reported by state and decade in the United States from 1944 to 1993 (total, 362 cases), showing increasing numbers of cases, increasing number of states reporting cases, and an eastward shift in state of occurrence (Centers for Disease Control data)

**Fig. 5.6** The Oriental rat flea, *X. cheopis* (from "Fleas of Alabama," Auburn University Agricultural Experiment Station, with permission)

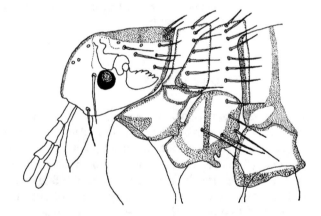

can develop both bubonic and pneumonic plague. Human cases associated with cats have occurred in seven states and have resulted in four deaths [6]. Infected cats may transmit plague organisms by direct contact through scratches or exudates from infected sores or even inhalation of infectious aerosols from the cats.

## 5.2.4  *Diagnosis and Treatment*

Plague should be considered in the differential diagnosis for any mysterious febrile illness occurring west of the Mississippi River (or with travel history out West) during the summer months. There may or may not be a history of flea bites. Questions

**Fig. 5.7** The ground squirrel flea, *Oropsylla montana* (CDC photo by John Montenieri)

about rodents living in close proximity to the home may be useful in determining exposure. Information about recent rodent die-offs around the home is especially indicative of plague exposure. If a bubo is present (Fig. 5.8), a milliliter of sterile saline can be injected into it and immediately aspirated back into the syringe. This fluid can be stained by Gram's stain and/or cultured. Blood and other body fluids can be stained and cultured. However, therapy should never be delayed or withheld because of a negative stain. Characteristically, the bacterium of plague is ovoid and has a distinctive "safety pin" appearance when stained with Giemsa, Wayson, or Wright's stain. Polymerase chain reaction (PCR), indirect fluorescent antibody (IFA) tests, and an antigen-capture enzyme-linked immunosorbent assay (ELISA) are available in some laboratories, permitting an early rapid diagnosis in acute cases [11]. Streptomycin is the drug of choice for the treatment of plague, although gentamicin or tetracycline may be satisfactory alternatives [11].

## 5.3  Murine Typhus

### 5.3.1  Introduction and Medical Significance

Murine typhus is a rickettsial disease transmitted to humans by fleas. The term "murine," of course, indicates that the disease is related to rats. In fact, the classic cycle involves rat-to-rat transmission with the Oriental rat flea, *X. cheopis*, being the main vector. Murine typhus is one of the most widely distributed arthropod-borne infections endemic in many coastal areas and ports throughout the world [12]. Outbreaks have been reported from Australia, China, Greece, Israel, Kuwait, and Thailand. At one time, there were thousands of cases reported annually in the United States; from 1931 to 1946, ~ 42,000 cases were reported [3, 13]. After World War

**Fig. 5.8** Plague bubo in right axilla of a human case (Armed Forces Institute of Pathology Neg. No. 219900[7B])

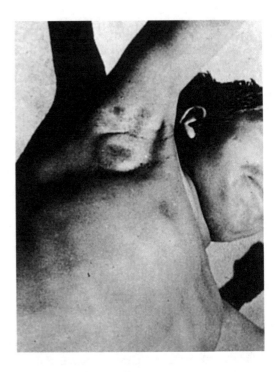

II, case numbers in the United States fell drastically to approximately 100 per year but seem to be rising again [14, 15]. A recent paper reported 90 cases of murine typhus from just 2 Texas hospitals over a 3-year period [16]. Most cases in the United States now are focused in central and south central Texas and Los Angeles and Orange counties in California [17]; however, physicians may encounter murine typhus in returning international travelers. Interestingly, the ecology of this disease seems to be changing. The classic rat–flea–rat cycle seems to have been replaced in some areas by a peridomestic animal cycle involving free-ranging cats, dogs, opossums, and their fleas.

## 5.3.2   Clinical and Laboratory Findings

Diagnosis of murine typhus is usually based on clinical suspicion. After an incubation period of 6–14 d, clinical symptoms may appear, including headache, chills, prostration, fever, and general pains. There may be a macular rash, especially on the trunk. The disease is usually mild with negligible mortality, except in the elderly. Severe cases occasionally occur with hepatic and renal dysfunction, central nervous system (CNS) abnormalities, and pulmonary compromise. Several points are useful to differentiate murine typhus from Rocky Mountain spotted fever (RMSF): RMSF mostly occurs in rural areas in the central, eastern, and southeastern United States (especially Oklahoma, Tennessee, and North Carolina). Murine typhus mostly occurs in urban or suburban areas in south Texas or southern California. RMSF patients often have a history of tick bite—or at least a history of exposure to tick-infested areas. Murine typhus patients often live in rat-infested buildings. The RMSF rash usually begins on the extremities and then moves to the trunk. Murine typhus rash begins on the trunk. (Note: These are just general guidelines; there are exceptions to each of these points.) Up to half of murine typhus patients have early mild leukopenia during the first 7 d of illness. Mildly elevated serum aspartate ami-notransferase levels are seen in about 90% of cases. Other fairly common lab find-ings are hypoalbuminemia and hypoproteinemia. IFA tests using specific *Rickettsia typhi* antigens, latex agglutination (LA) tests, and PCR are commonly used as diag-nostic tools for murine typhus infection. Diagnosis may also be established by com-plement fixation, but this test is old and generally unavailable. Since all typhus group *Rickettsiae* share common antigens, IFA tests may not discriminate between louse-borne and murine typhus unless the sera are differentially absorbed with the respective rickettsial antigen prior to testing. Antibody tests usually become posi-tive in the second week.

### 5.3.3 Ecology of Murine Typhus

There are numerous species of fleas, many of which are host-specific, feeding only on a certain animal. People often erroneously think "a flea is a flea" and that all spe-cies are equally important in disease transmission. Important fleas in disease cycles, such as murine typhus, are those that either (1) transmit the disease agent among the reservoir hosts (in this case, rats) or (2) transmit the agent to humans. The classic cycle of murine typhus in nature is as follows (nontypical cycles occur; *see* last paragraph): murine typhus is found in port areas of many parts of the world where the causative agent, *R. typhi*, is transmitted among domestic rats by fleas—primarily the Oriental rat flea, *X. cheopis* (Fig. 5.6). Other rat-feeding flea species may be involved, such as *Nosopsyllus fasciatus* and *Leptopsylla segnis*. Transmission may also occur among rats by a rat louse and/or a rat mite.

Distinction should also be made here about the rats involved. There are numer-ous species of rats—some considered "domestic" and others considered "wild." Wild rats include cotton rats, wood rats, and rice rats. Domestic rats, also known as commensal rats, are the species *Rattus norvegicus* and *R. rattus*. Their common names are the Norway rat and the black rat (or roof rat), respectively. In fleas, the pathogen generally proliferates in abundance within epithelial cells of the midgut, and when packed, these cells burst, releasing *Rickettsiae* into the lumen. Transmission to humans occurs when infective flea feces are scratched into the bite site (or other fresh skin wounds) or transported manually to the eyes or mucous membranes. There is some evidence that *R. typhi* may also be transmitted by flea bites, and not merely through contact with infective feces or crushed fleas [18].

In some areas of the world, murine typhus occurs in places where infected domestic rats and their fleas are absent. For example, in the United States, most cases of murine typhus occur in central and south central Texas and Los Angeles and Orange Counties in California. In those areas, infected rats and their fleas are hard to find. Studies have shown an abundance of opossums, cats, and dogs in these areas, but not rats. Also, the predominant flea on those animals is the cat flea, *Ctenocephalides felis*. (Cat fleas do not just feed on cats!) Azad and colleagues [12, 19] conducted several surveys in the areas and found cat fleas taken from opossums infected with both *Rickettsia typhi* (the causative agent of murine typhus) and *R. felis* (a relatively new typhus-like rickettsia). In addition, opossums and cats showed evidence of infection [20]. These findings indicate that the classic rat–flea–rat cycle of *R. typhi* has been replaced by a peridomestic animal cycle involving free-ranging cats, dogs, opossums, and their fleas (Fig. 5.9). This cycle is of potential public health importance since opossums and cat fleas are widespread pests over much of the United States—even in well-kept, upscale suburban neighborhoods.

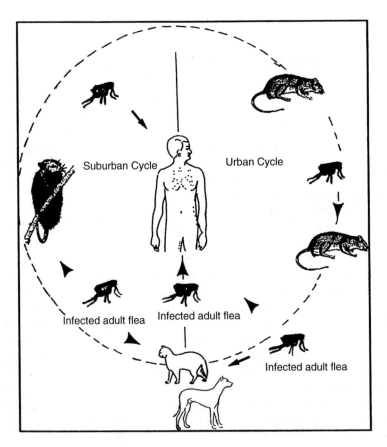

**Fig. 5.9** Urban and suburban life cycles of murine typhus (From CDC publication, Ref. 12)

More recently, *Rickettsia felis* infection has been found associated with fever, headache, myalgia, and macular rash in humans and has been detected in cat fleas in many places around the world [21, 22]. As mentioned, this agent is apparently maintained in nature in an opossum–cat flea cycle which is actually more common than the *R. typhi*–rat cycle in some areas.

### 5.3.4   Treatment

Drugs of choice for treatment of murine typhus include tetracycline, doxycycline, and chloramphenicol. Severely ill patients may require intravenous therapy. Antimicrobial therapy should be continued until 2–3 d after defervescence. Prevention and control of murine typhus are primarily directed toward control of the flea vectors and their animal hosts.

## 5.4   Cat-Scratch Disease (CSD)

### 5.4.1   Introduction and Clinical Presentation

The link between human cases of CSD and fleas is not firmly established; however, fleas certainly are involved in the natural history of the infection in animals. For this reason, CSD is included in this section. In humans, CSD is a subacute, usually self-limiting bacterial disease characterized by malaise, granulomatous lymphadenitis, and variable patterns of fever. The disease occurs following cat contact in 95–99% of patients, and the primary lesion of CSD evolves with development of papules, vesicles, and pustules at the inoculation site after 3–10 d [23]. Lymphadenopathy is common in lymph nodes draining the site of inoculation. In fact, CSD is a leading cause of subacute and chronic lymphadenopathy [24]. Although lymphadenopathy is the presenting sign in a majority of children, 25% of patients have atypical presentations, including Parinaud's oculoglandular syndrome, maculopapular rash, erythema nodosum, thrombocytopenic purpura, and encephalopathy [25]. Seizures and death may even occur as a result of CSD [26]. CSD is caused by *Bartonella (=Rochalimaea) henselae*, the most commonly recognized *Bartonella* infection in humans. It occurs worldwide, and there are approximately 13,000 cases reported in the United States each year [27]. Children and young adults are more often affected than older persons.

### 5.4.2   Reservoirs and Mode(s) of Transmission

As the name implies, CSD is carried by domestic cats. More than 90% of patients give a history of scratch, bite, lick, or other exposure to a healthy, usually young cat. However, dogs may also be involved [28]. At least one case was linked to exposure to a puppy [29]. Epidemiological studies on the risk factors associated with CSD have established a possible role for fleas in the transmission of *B. hense-lae* [30]. The organism has been detected in cat fleas by PCR techniques [31]. Higgins et al. [32] demonstrated that cat fleas can maintain *B. henselae* and excrete viable organisms in their feces for up to 9 d after feeding on an infected blood meal. Further experiments by Foil et al. [33] have shown that transmission of the CSD agent to humans could possibly be by flea bite but is most likely by exposure to infective flea feces. If this turns out to be the case, then fleas are only "mechanical" transmitters of the disease organism (*see* Chap. 2 for a discussion of mechanical vs biological transmission).

### 5.4.3   Treatment

CSD is usually self-limiting. In fact, treatment of typical CSD with antibiotics is controversial, since it may not alter the course of the disease [11, 24]. Antibiotics are, however, recommended for immunocompromised patients. The most commonly used antibiotic for CSD is azithromycin [11].

## References

1. CDC. Human plague – United States, 1993–1994. MMWR. 1994;43:242–3.
2. CDC. Summary of notifiable infectious diseases and conditions – United States, 2015. CDC, MMWR. 2017;64(53):1–144.
3. Harwood RF, James MT. Entomology in human and animal health. 7th ed. New York: Macmillan; 1979. 548 p
4. Crook L. Plague. In: Rakel RE, editor. Conn's current therapy. Philadelphia: W.B. Saunders Co.; 1997. p. 127–8.
5. Reynolds PC, Brown TL. Current trends in plague. Wing Beats (Fl Mosq Contr Assoc). 1997;8:8–10.
6. Anonymous. Plague. U.S. Army Publ. CHPPM Today, Aberdeen Proving Ground, MD, July issue; 1997. p. 14.
7. Mead PS. Plague. In: Guerrant RL, Walker DH, Weller PF, editors. Tropical infectious diseases: principles, pathogens, and practice. 3rd ed. New York: Saunders Elsevier; 2011. p. 276–84.
8. Mullett CF. The Bubonic Plague and England. Lexington: University of Kentucky Press; 1956. 402 p
9. Goddard J. Fleas and plague. Inf Med. 1999;16:21–3.
10. Hinnebusch BJ, Bland DM, Bosio CF, Jarrett CO. Comparative ability of *Oropsylla montana* and *Xenopsylla cheopis* fleas to transmit *Yersinia pestis* by two different mechanisms. PLoS Negl Trop Dis. 2017;11(1):e0005276.
11. Heymann DL, editor. Control of communicable diseases manual. 20th ed. Washington, DC: American Public Health Association; 2015.
12. Azad AF, Radulovic S, Higgins JA, Noden BH, Troyer JM. Flea-borne rickettsioses: ecologic considerations. Emerg Infect Dis. 1997;3:319–27.
13. Traub R, Wisseman CL, Azad AF. The ecology of murine typhus: a critcal review. Trop Dis Bull. 1978;75:237–317.

14. Blanton LS, Vohra RF, Bouyer D, Walker DH. Reemergence of murine typhus in Galvaston, Texas, USA, 2013. Emerg Infect Dis. 2015;21:484–6.
15. Civen R, Ngo V. Murine typhus: an unrecognized suburban vectorborne disease. Clin Infect Dis. 2008;46:913–8.
16. Afzal Z, Kallumadanda S, Wang F, Hemmige V, Musher D. Acute febrile illness and complications due to murine typhus, Texas, USA. Emerg Infect Dis. 2017;23:1268–73.
17. Purcell K, Fergie J, Richman K, Rocha L. Murine typhus in children, South Texas. Emerg Infect Dis. 2007;13:926–7.
18. Azad AF, Traub R. Transmission of murine typhus rickettsiae by *Xenopsylla cheopis* with notes on experimental infection and effects of temperature. Am J Trop Med Hyg. 1985;34:555–63.
19. Schriefer ME, Sacci JB Jr, Taylor JP, Higgins JA, Azad AF. Murine typhus: updated roles of multiple urban components and a second typhus-like rickettsia. J Med Entomol. 1994;31:681–5.
20. Sorvillo FJ, Gondo B, Emmons R. A suburban focus of endemic typhus in Los Angeles County: association with seropositive cats and oppossums. Am J Trop Med Hyg. 1993;48:269–73.
21. Hawley JR, Shaw SE, Lappin MR. Prevalence of *Rickettsia felis* DNA in the blood of cats and their fleas in the United States. J Feline Med Surg. 2007;9(3):258–62.
22. Wiggers RJ, Martin MC, Bouyer D. Rickettsia felis infection rates in an east Texas population. Texas Med. 2005;101(2):56–8.
23. Koehler JE. *Bartonella*: an emerging human pathogen. In: Scheld WM, Armstrong D, Hughes JM, editors. Emerging infections, Part I. Washington, DC: ASM Press; 1998. p. 147–63.
24. Easley RB, Cooperstock MS, Tobias JD. Cat-scratch disease causing status epilepticus in children. South Med J. 1999;92:73–6.
25. Smith RA, Scott B, Beverley DW, Lyon F, Taylor R. Encephalopathy with retinitis due to cat-scratch disease. Dev Med Child Neurol. 2007;49(12):931–4.
26. Fouch B, Coventry S. A case of fatal disseminated *Bartonella henselae* infection (cat-scratch disease) with encephalitis. Arch Pathol Lab Med. 2007;131(10):1591–4.
27. Nelson CA, Saha S, Mead PS. Cat-scratch disease in the United States, 2005–2013. Emerg Infect Dis. 2016;22(10):1741–6.
28. Chen TC, Lin WR, Lu PL, Lin CY, Chen YH. Cat scratch disease from a domestic dog. J Formos Med Assoc. 2007;106(2 Suppl):S65–8.
29. Tsukahara M, Tsuneoka H, Lino H, Ohno K, Murano I. *Bartonella henselae* infection from a dog. Lancet. 1998;352(9141):1682.
30. Zangwill KM, Hamilton DH, Perkins BA. Cat scratch disease in Connecticut: epidemiology, risk factors, and evaluation of a new diagnostic test. N Engl J Med. 1993;329:8–13.
31. Anderson B, Sims K, Regnery R, Robinson L, Schmidt MJ, Goral S, et al. Detection of *Rochalimaea henselae* DNA in specimens from cat scratch disease patients by PCR. J Clin Microbiol. 1994;32:942–8.
32. Higgins JA, Radulovic S, Jaworski DC, Azad AF. Acquisition of the cat scratch disease agent *Bartonella henselae* by cat fleas (Siphonaptera: Pulicidae). J Med Entomol. 1996;33:490–5.
33. Foil L, Andress E, Freeland RL, Roy AF, Rutledge R, Triche PC, et al. Experimental infection of domestic cats with *Bartonella henselae* by inoculation of *Ctenocephalides felis* (Siphonaptera: Pulicidae) feces. J Med Entomol. 1998;35:625–8.

# Chapter 6
# Sand Fly-Transmitted Diseases

## 6.1 Basic Sand Fly Biology

Sand flies are tiny gnats (Fig. 6.1) that breed in dark, moist areas with plenty of available organic matter, which serves as food for the larvae. Examples of breeding sites include hollow trees, animal burrows, and under dead leaves. Female sand flies have piercing mouthparts and are bloodsuckers. Males take moisture from any available source and are even said to consume human sweat. After a blood meal, the female scatters between 30 and 70 eggs in the potential breeding site; they hatch about 1–2 wk later. There are four larval stages, with each stage eating decaying organic matter and perhaps microorganisms. The pupal stage is inactive, and emergence occurs in 5–10 d. Adults seek out cool, moist places to rest, such as caves, cracks in rocks, or tree holes. At night they come out to feed. Many species prefer to feed on mammals, though some prefer reptiles and amphibians.

## 6.2 Leishmaniasis

### 6.2.1 Introduction and Medical Significance

Leishmaniasis is a term used to describe any one of several diseases caused by protozoan parasites in the genus *Leishmania*. It is a highly complex disease group with many contributing factors and unknowns; this chapter is an effort to present a simplified synopsis of what is known about the subject. For greater detail, the reader should consult other references [1–6]. Leishmaniasis is a sand fly-transmitted disease that occurs in almost all countries of the New World (especially tropical areas) and in many countries of the Old World, especially the areas surrounding the Mediterranean basin (Figs. 6.2 and 6.3). There is great diversity in ecological settings where leishmaniasis may occur—arid, rural areas, tropical forests, subalpine valleys, and even urban environments [7].

© Springer International Publishing AG, part of Springer Nature 2018
J. Goddard, *Infectious Diseases and Arthropods*, Infectious Disease,
https://doi.org/10.1007/978-3-319-75874-9_6

**Fig. 6.1** Adult sand fly
(Armed Forces Pest
Management Board)

**Fig. 6.2** Approximate geographic distribution of cutaneous leishmaniasis

Clinically, leishmaniasis manifests itself in four main forms: cutaneous, muco-cutaneous, diffuse cutaneous, and visceral [8]. The cutaneous form may appear as small and self-limited ulcers that are slow to heal. When there is destruction of nasal and oral mucosa, the disease is labeled mucocutaneous leishmaniasis. Sometimes there are widespread cutaneous papules or nodules all over the body—a condition termed diffuse cutaneous leishmaniasis. Finally, the condition in which the para-sites invade cells of the spleen, bone marrow, and liver—causing widespread vis-ceral involvement—is termed visceral leishmaniasis. There is much morbidity and

**Fig. 6.3** Approximate geographic distribution of visceral and mucocutaneous leishmaniasis

mortality owing to leishmaniasis worldwide: collectively, the leishmaniases are endemic in 97 countries with an estimated worldwide annual incidence of 700,000–1,000,000 cases and 20,000–30,000 deaths [9, 10]. Except for the possibility of seeing a cutaneous case in Texas, physicians in the United States will generally only see leishmaniasis in travelers, expatriates, immigrants, and returning soldiers. Several hundred cases of cutaneous leishmaniasis have been reported among US military personnel serving in Iraq.

## 6.2.2 Clinical Manifestations and Diagnosis

### 6.2.2.1 Old World Forms of Leishmaniasis

The classic form of Old World *cutaneous leishmaniasis*, called oriental sore, is most frequently caused by *Leishmania major, Leishmania tropica, Leishmania infantum*, or *Leishmania aethiopica*. After an incubation period of 2 wk to several months, a papule develops at the site where promastigotes (one of the stages of the parasite) were inoculated by the sand fly bite. The papule then gradually increases in size, becomes crusted, and ulcerates. The ulcer is often circular and shallow with raised, well-defined erythematous borders, resembling a pizza. Smaller lesions may look like buttons, e.g., the so-called Jericho buttons (Fig. 6.4). There may or may not be a serious discharge. Cutaneous lesions are slow to heal and may be accompanied by regional lymphadenopathy. Generally, after several months or a year or so, ulcers

**Fig. 6.4** "Jericho buttons"
(cutaneous leishmaniasis
lesions), known for
frequency of cases near the
ancient city of Jericho
(Photo courtesy from the
Matson Photograph
Collection, United States
Library of Congress,
public domain)

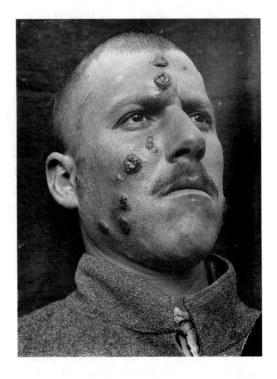

heal, leaving flat, atrophic, depigmented scars. A variant of cutaneous leishmania-sis, called diffuse cutaneous leishmaniasis, may begin as a papule but does not ulcerate. Other lesions such as plaques may form in the general vicinity of the initial papule and may also develop in distant areas of the body but especially the face and extremities. Diffuse cutaneous leishmaniasis may persist for 20 yr or more [11]. *Visceral leishmaniasis* (sometimes also called kala-azar) is most often caused by *Leishmania donovani* or *Leishmania infantum* and may be fatal if not treated. Patients generally display fever, splenomegaly, hepatomegaly, anemia, leukopenia, hypergammaglobulinemia, and weight loss. On the other hand, there can be mild cases with asymptomatic, self-resolving visceral infections. Young children and possibly malnourished populations seem to have a greater likelihood of developing visceral disease. Oddly, some foxhounds in the United States have been found infected with *L. infantum* in certain areas [12]. One survey found 12% of 11,000 foxhounds in the eastern United States with antibodies to the agent of visceral leish-maniasis. The significance of this finding is yet to be determined but indicates the potential for human infection in the United States.

### 6.2.2.2  New World Forms of Leishmaniasis

New World cutaneous leishmaniasis is predominantly caused by *Leishmania brazil-iensis*, *Leishmania guyanensis*, *Leishmania panamensis*, *and Leishmania mexi-cana.* The disease has been called pian bois (bush yaws), uta, and chiclero's ulcer.

**Fig. 6.5** Mucocutaneous leishmaniasis (Armed Forces Institute of Pathology, Neg. No. 74–8873-1)

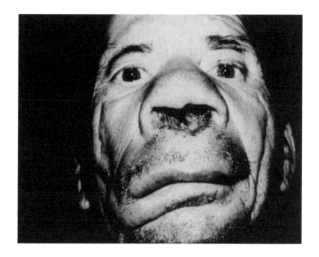

There can be single, localized ulcers that are slow to heal (months to years) or diffuse cutaneous forms that may resemble lepromatous leprosy. *Mucocutaneous leishmaniasis*, also known as espundia, develops in <5% of patients, typically after months or years, and usually follows cases of cutaneous leishmaniasis caused by *L. braziliensis* or *L. panamensis*. It is believed that localization in the nasal mucosa occurs during parasitemia associated with the initial infection. The disease may be severely disfiguring, eroding the cartilaginous tissues of the nose and palate (Fig. 6.5). Mucosal lesions never heal spontaneously, and death from secondary infections is common in untreated patients. There are sporadic cases of visceral leishmaniasis in South America usually caused by *L. infantum chagasi*. There have been outbreaks in Teresina and Natal, Brazil. Children are most frequently affected. The clinical picture of visceral leishmaniasis is a pentad of chronic fever, wasting, marked hepatosplenomegaly, pancytopenia, and hypergammaglobulinemia. Protean clinical manifestations may occur, especially early in the disease. Infection is often fatal if untreated but is usually <5% with adequate drug therapy.

For a long time, the abovementioned Old World and New World disease forms were correlated with various *Leishmania* species and geographic regions to make a well-defined, neat classification. As is the case in many paradigms in science, this classification is turning out not to be so clear-cut. There is apparently a whole spectrum of diseases—from cutaneous to visceral—depending on many factors, such as species of *Leishmania*, numbers of parasites (parasite burden), and the predominant host immune response [3]. The idea that a few *Leishmania* species each cause a distinct and separate clinical syndrome is no longer valid. For example, a visceral species, *L. chagasi*, has also been isolated from patients with cutaneous leishmaniasis in several Central American countries. However, the particular parasite species and geographic location may still serve as useful epidemiologic "labels" for the study of the disease complex, and generalized statements can be made. For example, *L. donovani* and *L. infantum* generally cause visceral leishmaniasis in the Old World, whereas *L. tropica* and *L. major* cause cutaneous lesions. In the New World, visceral disease is mainly caused by *L. infantum chagasi*, cutaneous lesions by *L. mexicana* and related species, and mucocutaneous lesions by *L. braziliensis*. Further, by sorting out which species do not cause human disease in a given area, researchers are better able to focus their studies on the ecology and behavior of those that do.

### 6.2.2.3   Diagnosis

Diagnosis of leishmaniasis is complicated because of the various forms of the disease (cutaneous, visceral, mucocutaneous), variety of parasite species involved, geographic variations, and other clinically similar syndromes. For example, blastomycosis, yaws, cutaneous tuberculosis, and other skin diseases may look like cutaneous leishmaniasis, but visceral leishmaniasis may be confused with malaria, typhoid fever, typhus, and schistosomiasis. Laboratory findings may be useful, especially in visceral leishmaniasis. There are usually anemia, leukopenia, and hypergammaglobulinemia. White blood cell (WBC) counts may occasionally be below 1000 per mm$^3$. Also, the ratio of globulin to albumin is typically high. The leishmanin skin test is sometimes used, although its results may be ambivalent. (However, the leishmanin test is mostly positive in cutaneous cases.) Serological tests, such as enzyme-linked immunosorbent assay (ELISA), agglutination assays, and indirect fluorescent antibody (IFA), are often employed (if appropriate antigens are available) to aid in diagnosis of leishmaniasis, especially visceral leishmaniasis. A more definite traditional "parasitological" diagnosis requires the demonstration of promastigotes in in vitro culture or amastigotes in Giemsa-stained histopathologic sections or smears from tissue aspirates. To look for parasites in cutaneous lesions, scrapings may be taken from the base of the ulcer or punch biopsies from the edge of suspicious skin lesions. More recently, PCR-/DNA-based assays are assuming a greater role in diagnosis of leishmaniasis [3].

## 6.2.3   Ecology of Leishmaniasis

*Leishmania* parasites are transmitted by female sand flies. In general, sand flies in the genus *Lutzomyia* are vectors in the Americas, while *Phlebotomus* are vectors everywhere else [3]. Infected flies transmit the flagellated form, called promastigotes, to a mammal host when taking a blood meal (Fig. 6.6). Promastigotes enter monocytes or macrophages and subsequently transform into oval amastigotes. Infected macrophages may later be ingested by feeding sand flies, thus completing the cycle. Sand flies are in the family Psychodidae. They belong to a particular subfamily, the Phlebotominae, which have piercing mouthparts and are bloodsuckers. Many species feed on cold-blooded animals, such as lizards, snakes, and amphibians; others feed on a variety of warm-blooded animals, including humans [13]. As is the case with mosquitoes, the females require a blood meal for ovarian development; males take moisture from a variety of sources. Sand flies are weak fliers, mostly active at night (a few species are day biters) and usually only when there is little or no wind.

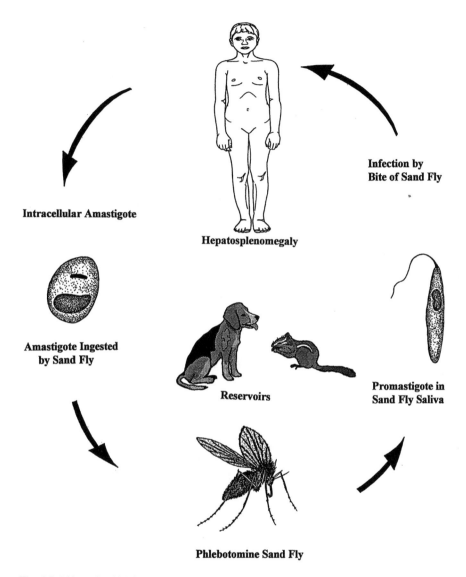

**Fig. 6.6** Life cycle of leishmaniasis (Provided by permission from Ref. 5; redrawn originally from Ref. 6)

### 6.2.3.1   Indigenous Leishmaniasis in the United States

There is a region primarily in southern Texas (roughly San Antonio south to the Mexican border) in which locally acquired cases of cutaneous leishmaniasis occur. There have been approximately 50 human cases reported from this region since the 1970s, which generally presented as slow-to-heal ulcers on body parts exposed to sand fly bites [14]. McHugh and colleagues [7, 15–17] unraveled the complex ecology/life cycle of leishmaniasis in that area, finding that the enzootic cycle involves transmission among wood rats (*Neotoma*) by the sand fly, *Lutzomyia anthophora*, which inhabits nests of rodents. Humans apparently become infected when they live near or are active in the cactus–mesquite habitat of wood rats. Because *L. anthophora* does not commonly feed on humans, a second sand fly species may act as a bridge from wood rats to humans. Thus far, all parasite isolates from humans, sand flies, rodents, and a single cat infection in Texas have been identified as *L. mexicana*, a relatively benign species. The contribution of field researchers and others in this area is critically important from the public health standpoint. How can physicians, epidemiologists, and public health workers make recommendations to the public for the prevention and control of a disease in an area, if nothing is known of the causative agent, its reservoir host(s), and its vectors? Much basic research is still needed (even in the "modern" and developed United States) in the vector-borne and parasitic diseases.

### 6.2.4  Treatment and Control of Leishmaniasis

The drugs of first choice for treatment of leishmaniasis have traditionally been the pentavalent antimonials, Glucantime and Pentostam. However, development of resistance and reports of substantial negative side effects have limited their use in some areas. In those cases, other drugs have been used, such as liposomal amphotericin B. Sometimes visceral leishmaniasis patients are treated with a combination of $\gamma$-interferon and pentavalent antimony. The Parasitic Disease Drug Service and Centers for Disease Control should be consulted for the most up-to-date information on treatment strategies for leishmaniasis. Unfortunately, infections are often chronic and can recur if/when the patient's immune system is suppressed [18]. Prevention of leishmaniasis is mostly limited to educating travelers about the risks of leishmaniasis and avoiding sand fly bites. Repellents containing the active ingredient DEET can provide limited, partial protection. Use of fine mesh bed nets impregnated with permethrin can dramatically reduce the number of sand fly bites. In addition, insecticides sprayed around human habitations can be useful in protecting people from leishmaniasis, especially sand fly species that are peridomestic. Other successful strategies include the selective use of poisoned baits or traps for animal reservoirs and environmental modifications, such as localized clearing of forests, subsoil plowing to destroy burrows, and vegetation modifications.

## 6.3  Other Sand Fly-Transmitted Diseases

### 6.3.1  Bartonellosis (Carrion's Disease)

Bartonellosis is a bacterial infection caused by *Bartonella bacilliformis*, which occurs in the Andes Mountains in parts of Peru, Ecuador, and southwest Colombia. Vector sand flies are *Lutzomyia verrucarum* and *Lutzomyia colombiana*. There are two clinical forms of the disease: a febrile anemia (Oroya fever) and a benign dermal eruption (verruga peruana) [19], but both forms are still considered Carrion's disease [20]. Oroya fever, the acute stage, occurs after an incubation period of about 3 wk. Clinical signs and symptoms include lymphadenopathy, hepatosplenomegaly, and anemia (often severe) [21]. The mortality rate is about 8%, with most dying of acute anemia. Several months after the resolution of Oroya fever, many patients develop verruga peruana (Peruvian warts). The verrugae are chronic, lasting from several months to years, and contain large numbers of *B. bacilliformis* bacilli. Antibiotic treatment can slow the lysis of erythrocytes in Oroya fever but may not prevent the subsequent development of verrugae [22].

## 6.3.2    Sandfly Fever

Sandfly fever (phlebotomus fever, papatasi fever, three-day fever) is caused by at least three distinct virus serotypes resulting in a febrile illness in humans lasting from 2 to 4 d or even longer. There may be accompanying retrobulbar pain on motion of the eyes, malaise, nausea, and pain in the limbs and back [23]. Symptoms may be alarming, but death is rare. This group of viruses occurs mostly in Europe, Asia, and Africa, although there is at least one in Central and South America. The classic "sandfly fever" is common during the summer months throughout the Mediterranean basin, the Middle East, Pakistan, and parts of India and Central Asia. The vector of the classic form throughout the entire Old World is thought to be the common sand fly, *Phlebotomus papatasi*. Other sand fly species such as *P. perniciosus* and *P. perfiliewi* are vectors of the other main serotypes.

## References

1. Ashford R, Bettini S. Ecology and epidemiology: old world. In: Peters W, Killick-Kendrick R, editors. The leishmaniases in biology and medicine. London: Academic Press; 1987. p. 365–420.
2. Chang K, Bray R. Leishmaniasis. Amsterdam: Elsevier Science Publishers; 1985. p. 986.
3. Jeronimo SMB, Sousa AQ, Pearson RD. Leishmaniasis. In: Guerrant RL, Walker DH, Weller PF, editors. Tropical infectious diseases: principles, pathogens, and practice. 3rd ed. New York: Saunders Elsevier; 2011. p. 696–706.
4. Magill AJ. Epidemiology of the leishmaniases. Dermatoepidemiol. 1995;13:505–23.
5. Goddard J. Leishmaniasis. Infect Med. 1999;16:566–9.
6. Markell E, Voge M, John D. Medical parasitology. 7th ed. Philadelphia: W.B. Saunders; 1992.
7. McHugh CP. Arthropods: vectors of disease agents. Lab Med. 1994;25:429–37.
8. Lane RP, Crosskey RW. Medical insects and arachnids. New York: Chapman and Hall; 1996. 723 p.
9. WHO. Leishmaniasis. World Health Organization, Global Health Observatory Data. 2017. http://www.who.int/gho/neglected_diseases/leishmaniasis/en/.
10. WHO. Media center – leishmaniasis. World Health Organization, Media Center Fact Sheet. 2017. http://www.who.int/mediacentre/factsheets/fs375/en/.
11. Pearson RD, Sousa ADQ. Leishmania species: visceral, cutaneous, and mucosal leishmaniasis. In: Mandell GL, Bennett JE, Dolin R, editors. Principles and practice of infectious diseases. 4th ed. New York: Churchill Livingstone; 1995. p. 2428–38.

12. Enserink M. Has leishmaniasis become endemic in the U.S.? Science (News Focus). 2000;290:1881–3.
13. Harwood RF, James MT. Entomology in human and animal health. 7th ed. New York: Macmillan; 1979. p. 548.
14. McHugh CP, Melby PC, LaFon SG. Leishmaniasis in Texas: epidemiology and clinical aspects of human cases. Am J Trop Med Hyg. 1996;55:547–55.
15. Kerr SF, McHugh CP, Dronen NOJ. Leishmaniasis in Texas: prevalence and seasonal transmission of *Leishmania mexicana* in *Neotoma micropus*. Am J Trop Med Hyg. 1995;53:73–7.
16. McHugh CP, Grogl M, Kreutzer RD. Isolation of *Leishmania mexicana* from *Lutzomyia anthophora* collected in Texas. J Med Entomol. 1993;30:631–3.
17. McHugh CP, Kerr SF. Isolation of *Leishmania mexicana* from *Neotoma micropus* collected in Texas. J Parasitol. 1990;76:741–2.
18. Golino A, Duncan JM, Zeluff B, DePriest J, McAllister HA. Leishmaniasis in a heart transplant patient. J Heart Lung Transplant. 1992;11:820–3.
19. Heymann DL, editor. Control of communicable diseases manual. 20th ed. Washington, DC: American Public Health Association; 2015.
20. Dehio C, Maguina C, Walker DH. Bartonelloses. In: Guerrant RL, Walker DH, Weller PF, editors. Tropical infectious diseases: principles, pathogens, and practice. 3rd ed. New York: Saunders Elsevier; 2011. p. 265–71.
21. Koehler JE. *Bartonella*: an emerging human pathogen. In: Scheld WM, Armstrong D, Hughes JM, editors. Emerging infections, Part I. Washington, DC: ASM Press; 1998. p. 147–63.
22. Weinman D, Kreier JP. *Bartonella* and *Grahamella*. In: Kreier JP, editor. Parasitic Protozoa, vol. 4. New York: Academic Press; 1977. p. 197–233.
23. Tesh RB, Vasconcelos PFC. Sandfly fever, Oropouche fever, and other Bunyavirus infections. In: Guerrant RL, Walker DH, Weller PF, editors. Tropical infectious diseases: principles, pathogens, and practice. 3rd ed. New York: Saunders Elsevier; 2011. p. 481–2.

# Chapter 7
# Miscellaneous Vector-Borne Diseases

## 7.1 Chagas' Disease

### 7.1.1 Introduction and Medical Significance

Chagas' disease, or American trypanosomiasis, is one of the most important arthropod-borne diseases in the Western Hemisphere. It mostly occurs in Mexico and Central and South America (Fig. 7.1), but at least 25 indigenous cases have been reported in the United States [1, 2]. This is likely an underestimate based on studies of vector blood meals and infection rates with the agent [3]. There are an estimated 6–8 million people with Chagas' disease worldwide and at least 100 million people at risk [2, 4]. Often being a long, chronic, and debilitating disease, Chagas' disease causes tremendous economic losses. The economic loss for South America alone owing to early mortality and disability in economically most productive young adults amounts to billions of dollars. Chagas' disease is caused by *Trypanosoma cruzi*, a protozoan that occurs in humans as a hemoflagellate and as an intracellular parasite without an external flagellum. Vectors of Chagas' disease are hemipteran insects (the true bugs) in the family Reduviidae, subfamily Triatominae. They are commonly called "kissing bugs" (Fig. 7.2) because of the nasty habit of taking a blood meal from around the lips of a sleeping victim. (However, this is an overgeneralization; the bugs will bite on exposed skin just about anywhere on the body.) Chagas' disease has both acute and chronic forms but is perhaps best known for its chronic sequelae, including myocardial damage with cardiac dilatation, arrhythmias and major conduction abnormalities, and digestive tract involvement, such as megaesophagus and megacolon [4–7].

© Springer International Publishing AG, part of Springer Nature 2018
J. Goddard, *Infectious Diseases and Arthropods*, Infectious Disease,
https://doi.org/10.1007/978-3-319-75874-9_7

**Kissing Bug Allergy**

Arthropod bites, as opposed to stings, may produce allergic reactions in humans, presumably a result of hypersensitivity to salivary components secreted during the biting process. These salivary secretions contain anticoagulants, enzymes, agglutinins, and mucopolysaccharides which may serve as sensitizing allergens. Reactions have occurred following bites by many different types of insects but most commonly from bites by *Triatoma* (kissing bugs), horse and deer flies, and mosquitoes. Kissing bugs—so named because of the nasty habit of taking a blood meal from the face—belong to the insect family Reduviidae (hence the sometimes used moniker "reduviid bugs") but, specifically, the subfamily Triatominae. Within this subfamily, some (but not all) species fall under the genus *Triatoma*—triatomines may also be in other genera. There are at least ten *Triatoma* species found in the United States, but only about six of these are likely to be encountered. Allergic reactions have been reported from bites by five species (*T. protracta, T. gerstaeckeri, T. sanguisuga, T. rubida*, and *T. rubrofasciata*), although in the United States, *T. protracta* is the species most often reported in allergic reactions. Kissing bug bites may be painless, leaving a small punctum without surrounding erythema, or cause delayed local reactions appearing like cellulitis. Anaphylactic reactions include itchy, burning sensations, respiratory difficulty, and other typical symptoms of anaphylaxis.

*Triatoma* bugs feed on vertebrate hosts such as bats, other small and medium-sized mammals, birds, and humans. Accordingly, the pests are often found in association with their host nest or habitat—caves, bird nests, rodent burrows, human houses, etc. For example, *T. protracta* is found in wood rat nests. Bugs periodically fly away from the nests of their hosts (nocturnal cyclical flights) and may be attracted to lights at dwellings, subsequently gain entrance and try to feed. Some species are able to colonize houses; they seem especially prolific in substandard structures with many cracks and crevices, mud walls, thatch roofs, etc.

Personal protection measures from kissing bugs involve avoidance (if possible)—such as not sleeping in adobe or thatched roof huts in endemic areas —and exclusion methods such as putting up bed nets. Domestic or peridomestic kissing bug species (Mexico, Central and South America) may be controlled by proper construction of houses, sensible selection of building materials, sealing of cracks and crevices, and precision targeting of insecticides within the home. In the United States, prevention of bug entry into homes may involve outdoor light management (i.e., lights placed away from the house, shining back toward it, instead of lights on the house) and efforts to find and seal entry points around the home.

**Fig. 7.1** Approximate geographic distribution of human cases of Chagas' disease

## *7.1.2   Clinical and Laboratory Findings*

Human infection with *T. cruzi* often leads to a chagoma (localized induration) at the site of infection. It is possible for similar lesions to appear subsequently anywhere on the body during the first few weeks of infection, presumably by hematogenous spread. When the conjunctiva is the route of entry, unilateral edema may appear, affecting both the upper and lower eyelid—known as Romaña's sign—which is a frequent characteristic of the acute stage [8] (Fig. 7.3). Note: Many patients with acute Chagas' disease develop neither a chagoma nor Romaña's sign. Other signs and symptoms of acute Chagas' disease include fever, malaise, lymphadenopathy,

**Fig. 7.2** Kissing bugs. (Courtesy Joe MacGown, Mississippi State University, used with permission)

**Fig. 7.3** Child showing Romaña's sign. (Armed Forces Institute of Pathology. Neg. No. 62-3934-6)

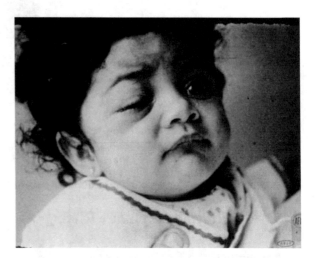

and hepatosplenomegaly. Less than 1% of patients with acute disease die owing to myocarditis and meningoencephalitis. Pathology during the acute stage is related to high parasitemias characterized by the presence of inflammatory infiltrates in several tissues, including heart and skeletal muscle, as well as by an increased production of inflammatory mediators, such as γ-interferon, tumor necrosis factor, interleukin 1, and oxygen and nitrogen reactive intermediates [9]. Patients surviving the acute stage often enter a symptomless phase lasting for months or years, during which time the parasites invade many organs of the body. This is sometimes called an "indeterminate phase" defined by the absence of clinical, radiological, and electrocardiographic manifestations of cardiac or digestive involvement in chronically infected persons. However, advanced cardiovascular tests may be able to identify significant abnormalities [10]. Moreover, patients with the indeterminate form have a poor prognosis: after 5–10 yr, a third of them will have cardiopathy [10]. Chronic Chagas' disease involves irreversible symptoms, such as arrhythmias, conduction blockage, aneurysms, myocarditis, megaesophagus, and megacolon. Many of these patients progressively become weaker and die from heart failure or other complications. Electrocardiograms (EKGs) of persons in the chronic stage are characteristically altered—most show partial or complete atrioventricular (AV) block, complete right bundle branch block, or premature ventricular contractions, along with abnormalities of the QRS complexes and of the P and T waves.

Diagnosis of Chagas' disease in the acute phase is made by demonstration of the trypomastigote stage of the parasites in peripheral blood (mainly), lymph node, or skeletal tissue. Parasitemia is most intense during the earliest stages of infection, so finding the parasites becomes increasingly difficult as time goes by. Complicating matters, there is a nonpathogenic trypanosome, *Trypanosoma rangeli*, infecting people in Brazil, Venezuela, Colombia, Panama, El Salvador, Costa Rica, and Guatemala, that must be differentiated from *T. cruzi*, but *T. rangeli* is longer than *T. cruzi*—*T. rangeli* about 30 μm and *T. cruzi* about 20 μm. In addition to direct observation of the parasites, *T. cruzi* can be cultured in selective media or demonstrated in animal tissues (such as after intracerebral inoculation of suckling mice). Where available, xenodiagnosis is a useful diagnostic tool—feeding uninfected *Triatoma* bugs on the patient and finding the parasite in the bug's feces or intestines several weeks later. PCR is also a tool for detection of *T. cruzi* infection. One study showed that the 220-bp amplified fragment (the E13 element) is specific for *T. cruzi* DNA and very useful to detect the presence of the parasite in the blood from chronic chagasic patients [11]. Immunodiagnostic tests include complement fixation, indirect hemagglutination, indirect fluorescent assay (IFA), radioimmunoassay (RIA), and enzyme-linked immunosorbent assay (ELISA). False-positives are a persistent problem with these conventional assays.

## 7.1.3 Ecology of Chagas' Disease and Its Vectors

Kissing bugs (also called "conenose bugs" because of their cone-shaped head and beak) are 1–3 cm long, are good fliers, and have a short three-segmented beak well-suited for sucking vertebrate blood (Figs. 7.4 and 7.5). Kissing bugs feed on humans, opossums, armadillos, rats, various carnivores, and monkeys. There are at least a hundred species of kissing bugs, all in the subfamily Triatominae, but not all are equally important as vectors of Chagas' disease (Table 7.1). Some have adapted to human environments and are called domestic species, whereas others are never or almost never in/around human dwellings and are called sylvatic species. Domestic species

**Fig. 7.4** Typical kissing bug. (Photo copyright 2008 by Jerome Goddard, PhD)

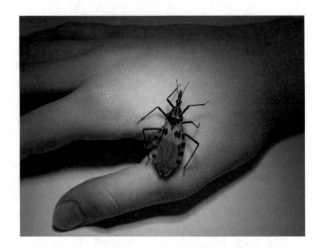

**Fig. 7.5** Three-segmented beak of the kissing bug

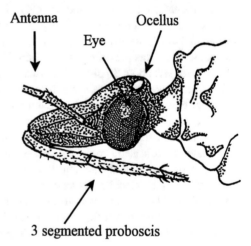

**Table 7.1** Five important vectors of Chagas' disease

| Species | Approximate distribution | Comments |
|---|---|---|
| *Triatoma infestans* | Argentina, Brazil, Bolivia, Chile, Paraguay, southern Peru, Uruguay | Highly adapted to the domestic environment; almost painless bite |
| *Panstrongylus megistus* | Argentina, Brazil, Paraguay | Almost painless bite |
| *Rhodnius prolixus* | Colombia, Costa Rica, El Salvador, French Guiana, Guatemala, Guyana, Honduras, parts of Mexico, Nicaragua, Venezuela | Principal vector in Venezuela and Colombia |
| *Triatoma brasiliensis* | Brazil | Almost painless bite |
| *Triatoma dimidiata* | Belize, Colombia, Costa Rica, Ecuador, El Salvador, Guatemala, Honduras, Mexico, Nicaragua, northern Peru, Venezuela | Almost painless bite |

are nocturnal; during daytime they seek refuge in the cracks and crevices in poorly constructed (often mud) houses or in the loose thatched roofing of huts. Interestingly, when insect control programs eliminate the domestic species, sylvatic species sometimes move in to take their place.

The causative agent of Chagas' disease is a flagellate protozoan, *T. cruzi*, which has a life cycle involving both a mammalian and a hemipteran insect host. In mammals, *T. cruzi* occurs in tissues in a nonflagellated form, called an amastigote, and in blood as a flagellated form, called a trypomastigote. In the bug, development is complicated, involving both metamorphic changes and multiplication; the parasites are eventually passed in the feces of the insect. Human infection does not occur by salivary transmission but, instead, through the feces of the bug, which almost always defecates on the skin of the victim while in the act of sucking blood. Patients may inadvertently rub or scratch fecal material into the bite wound. Infection may also be achieved through blood transfusion; *T. cruzi* has been shown to remain viable in refrigerated blood for at least several weeks [12]. Sexual or congenital transmission is also possible [13]. Other routes of transmission should not be overlooked. Contaminated sugarcane juice, acai fruit juice, or infected raw meat may be infective [14]. In some communities in Mexico, people believe that bug feces can cure warts or that the bugs have aphrodisiac powers [15]. In addition, Mexican children often play with triatomine bugs collected in their houses, and in Jalisco reduviid bugs are eaten with hot sauce by the Huichol Indians [15].

## 7.1.4 Treatment, Prevention, and Control

Treatment of Chagas' disease is problematic and controversial. In some countries allopurinol and itraconazole have been used for treatment of chronic disease in adults [16]. For acute Chagas' disease, two drugs, nifurtimox and benznidazole, are currently used, but early diagnosis is difficult and severe side effects can occur [5, 6]. Studies have shown that treatment of patients with either no or little evidence of heart disease is effective, but specific treatment (e.g., with benznidazole) for patients with advanced Chagas' disease is unlikely to help [17]. The likelihood of developing a safe vaccine for Chagas' disease is remote because *T. cruzi* antigens can stimulate autoimmune reactions. Therefore, control of Chagas' disease depends heavily upon interrupting parasite transmission by removing the vectors (disinfestation of houses) and screening blood banks for infected blood. One interinstitutional cooperative project in Latin America called the Southern Cone Initiative which involved mandatory blood screening and house fumigation led to an impressive reduction in *T. cruzi* infection [18]. Personal protection measures from kissing bugs involve avoidance (if possible)—not sleeping in adobe or thatched roof huts in endemic areas—and exclusion methods, such as bed nets. Prevention and control of domestic species of triatomines can be accomplished by proper construction of houses, wise choice of building materials, sealing cracks and crevices, and precision targeting of insecticides within the home.

**Fig. 7.6**  Adult tsetse fly

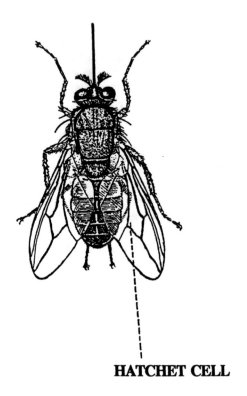

**HATCHET CELL**

## 7.2   African Sleeping Sickness

### 7.2.1   Introduction and Medical Significance

African sleeping sickness, also known as human African trypanosomiasis (HAT), is transmitted by tsetse flies (Fig. 7.6) and usually occurs at a low level of transmission in tropical Africa, with occasional epidemic outbreaks. The disease is caused by two closely related organisms, *Trypanosoma brucei gambiense* and *Trypanosoma brucei rhodesiense*. It is called sleeping sickness because there is often a steady progression of meningoencephalitis, with increase of apathy and somnolence. The patient may gradually become more and more difficult to arouse and finally becomes comatose. The *T.b. gambiense* form of the disease may run a protracted course of many years; the *T.b. rhodesiense* form can be lethal within weeks or a few months without treatment. Both forms are almost always fatal if untreated. Historically, HAT has been a major impediment to the social and economic development of Central and eastern Africa (Fig. 7.7). With the use of modern drugs and insecticides, this disease was effectively reduced in most countries by the mid-1960s [19] but rebounded in the late 1900s, mainly due to war disrupting control programs. Countries such as the Sudan, Democratic Republic of Congo, and Angola currently have the most problems with sleeping sickness. Fortunately, case numbers are declining, with only about 3000 reported in 2015 [20].

**Fig. 7.7** Approximate geographic distribution of African trypanosomiasis

## 7.2.2   Clinical and Laboratory Findings

In the early stages, HAT is characterized by fever, malaise, headache, and anorexia. The fever is usually irregular and may be initiated by a rigor. Night sweats are frequent. Often, the cervical lymph nodes enlarge, a condition called Winterbottom's sign. In the latter stages, there are body wasting, somnolence, and signs referable to the central nervous system (CNS) [5]. Diagnosis is made by demonstrating the trypanosomes in the blood (Fig. 7.8), cerebrospinal fluid (CSF), or the lymph. Antibodies, specific for *T. b. gambiense* or *T. b. rhodesiense*, may be demonstrated by ELISA, IFA, or agglutination tests; high levels of IgM are common in African trypanosomiasis [5]. Also, there is a fairly accurate card agglutination test (CATT) which can be performed in remote areas without electricity which gives results within 10 min [21].

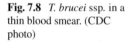

**Fig. 7.8** *T. brucei* ssp. in a thin blood smear. (CDC photo)

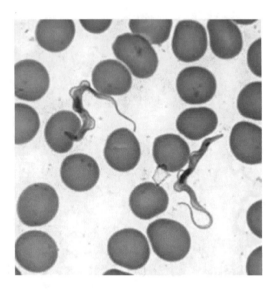

### 7.2.3   Ecology of African Sleeping Sickness and Its Vectors

There are over 20 species of tsetse flies in the genus *Glossina*. Tsetse flies feed on a wide variety of mammals and a few reptiles; people are not their preferred hosts. Both sexes feed on blood and bite during the day. Tsetse flies have a life-span of about 3 mo (less for males), and females give birth to full-grown larvae on dry and loose soil in places like thickets and sandy beaches. The larvae burrow a few centimeters into the substrate and pupate. The pupal stage lasts 2 wk. to 1 mo, after which the adult fly emerges to continue the life cycle.

Many species of tsetse are potential vectors of trypanosomes of people and animals; however, six species are of primary importance as vectors of HAT. Briefly, the chief vectors of *T. gambiense*, the cause of the Gambian form of sleeping sickness, are *Glossina palpalis*, *Glossina fuscipes*, and *Glossina tachinoides*. The Gambian form has humans, hogs, cattle, and sheep as reservoir hosts. Cases of Gambian sleeping sickness occur in western and central Africa and are usually more chronic. In eastern Africa, the Rhodesian form, which is more virulent, is caused by *T. rhodesiense*. The Rhodesian form has a number of wild animal reservoirs. The primary vectors of the Rhodesian form to humans are *Glossina morsitans*, *Glossina swynnertoni*, and *Glossina pallidipes*. Tsetse flies are generally confined to tropical Africa between 15°N and 15°S latitude. *Glossina morsitans* is a bush species found in wooded areas and brush country in eastern Africa. In western and central Africa, where members of the *G. palpalis* group are the principal vectors, the flies are predominantly found near the specialized vegetation lining the banks of streams, rivers, and lakes.

## 7.2.4   Treatment, Prevention, and Control

Treatment for HAT differs according to the form and phase of the disease. Generally, it is treated with pentamidine, although suramin, eflornithine, nifurtimox, and melarsoprol may also be used. However, melarsoprol may be fatal in 3–10% of patients treated [22, 23], and there are other complicating factors with all these drugs. Physicians attempting to treat a patient with HAT should seek the most up-to-date treatment recommendations from the WHO or the Parasitic Drug Service, Centers for Disease Control, Atlanta, GA. Large-scale control efforts include bush clearing along streams to control breeding sites, aerial spraying of insecticides, surveillance, case detection, and treatment of infected persons. Some recent success in vector control has been attained using novel fly trapping techniques (Fig. 7.9). In addition, the sterile male release technique has been used for tsetse fly control [23]. Personal protection measures against tsetse flies include wearing heavy-material, long-sleeved shirts and long pants, as well as screening, bed nets, and insect repellents.

## 7.3   Onchocerciasis

### 7.3.1   Introduction and Medical Significance

Onchocerciasis, caused by the filarial worm *Onchocerca volvulus*, is a nonfatal illness producing dermal nodules and ocular disease. Eye involvement may lead to blindness (called "river blindness"). The disease, occurring in sub-Saharan Africa and parts of Mexico and Central and South America (Fig. 7.10), is transmitted to humans by the bite of black flies (genus *Simulium*) (Fig. 7.11). The World Health Organization estimates that there are 17 million people with onchocerciasis and 1.1 million disability-adjusted life years lost [24]. Fortunately, onchocerciasis has been reduced in many countries in West Africa owing to intensive vector control programs and use of the antiparasitic drug, ivermectin. Until 1987, suramin and diethylcarbamazine were the only drugs available for the treatment of onchocerciasis, and they could not be used for community therapy because of their toxicity and the dosage schedules required [25]. The registration of Mectizan (ivermectin, MSD) for treatment of human onchocerciasis in 1987 and the donation of this drug by Merck and Company for as long as needed provided a new opportunity for the control of this disease. Ivermectin-based control through community-directed treatment has been introduced in 19 other endemic African countries through the African Program for Onchocerciasis Control (APOC) and in Yemen and South and Central America through the Program for the Elimination of Onchocerciasis in the Americas (OEPA) [26].

**Fig. 7.9**  Tsetse fly trap in Uganda. (Photo courtesy Rachel Freeman, RN)

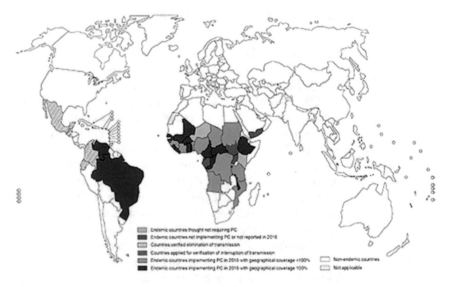

**Fig. 7.10**  Distribution of onchocerciasis. (World Health Organization figure)

## 7.3.2  Clinical and Laboratory Findings

Onchocerciasis is characterized by fibrous nodules in subcutaneous tissues. Adult filarial worms reside in these nodules. Microfilariae (baby worms) are constantly discharged from these nodules invading various tissues in the body, especially the

**Fig. 7.11** Black fly, vector of the filarial worm, *O. volvulus*. (Photo by Jerome Goddard and courtesy the Mississippi State University Extension Service)

skin and eyes. Microfilariae dying in the skin produce an intense pruritic rash, chronic dermatitis-altered pigmentation, edema, and atrophy [5]. Microfilariae reaching the eye may cause visual disturbances and/or blindness. Diagnosis is made by demonstrating microfilariae in skin biopsies or urine or by excising nodules and finding adult worms. Sometimes, ultrasound can be used to demonstrate adult worms in nodules [27]. In low-density infections, where microfilariae are not found in the skin and are not present in the eyes, the Mazzotti test (which can be dangerous in heavily infected patients) was used in the past for diagnosis, although PCR technology is now used to detect low-level infections.

### 7.3.3  Ecology of Onchocerciasis and Its Vectors

Black flies are small, stout-bodied flies that have blade-like mouthparts. They breed in swift running water (as opposed to mosquitoes, which breed in still water), such as streams and rivers, where the larvae attach themselves to rocks and other objects on the bottom. Adults emerge and generally fly within a range of 12–18 km (some much further) looking for food and mater. Females require a blood meal for egg development; the males never suck blood. Black flies occur in huge swarms, tormenting humans, and wild and domestic animals. They are particularly abundant in the north temperate and subarctic zones, but many species occur in the subtropics and tropics where factors other than seasonal temperatures affect their developmental and abundance patterns.

Not all black flies are vectors of onchocerciasis. In Africa, members of the *Simulium damnosum* and *Simulium neavei* complexes are important vectors. In fact, members of the *S. damnosum* complex are responsible for over 90% of onchocerciasis cases worldwide and more than 95% of cases in Africa [28]. In Central and

South America, vectors of onchocerciasis include *S. ochraceum* and *S. metallicum*. Humans are the definitive host for the parasite; there is no animal reservoir. Microfilariae are imbibed with the blood meal when female black flies feed on infected hosts. The tiny worms try to escape into the fly's hemocoel, but most fail to do so. The few microfilariae that succeed in breaking through the gut wall migrate to the flight muscles of the thorax where they undergo further development. Eventually, the parasites transform into active, third-stage larval worms, which move into the fly's head ready for transmission when it next bites [28].

### 7.3.4   Treatment, Prevention, and Control

Control of black flies involves application of insecticides for both adults and larvae. This has only limited success, since it is often difficult to locate and treat all breeding sites. Larviciding with the "biological" control agent, *Bacillus thuringiensis* (a spore-forming bacterium that kills the feeding larvae), has shown success in many African countries participating in the Onchocerciasis Control Program (OCP). The drug of choice for the management of onchocerciasis is ivermectin although doxycycline may be included with the ivermectin treatment [5, 29]. Ivermectin impairs the release of immature worms (microfilariae) from gravid females, thus reducing symptoms and transmission, but does not kill the adult worms. Treatment may be required for 10 yr or more until the natural death of adult worms.

## 7.4   Scrub Typhus (ST)

### 7.4.1   Introduction and Medical Significance

ST, a zoonotic rickettsial infection caused by *Orientia tsutsugamushi*, is mite-borne and occurs over much of Southeast Asia, India, Sri Lanka, Pakistan, islands of the southwest Pacific, and coastal Australia (Queensland) (Fig. 7.12). Recently, ST has been detected in the Middle East and South America, leading public health officials to worry about an ever-widening impact [30, 31]. The disease threatens over a billion people, and there are an estimated 1,000,000 cases per year [32]. Chiggers are the vectors, so the name "chigger-borne rickettsiosis" might be more appropriate. ST occurs in nature in small but intense foci of infected host animals. These "mite islands" or "typhus islands" occur where the appropriate combination of rickettsiae, vectors, and suitable animal hosts occurs [5, 33]. Epidemics occur when susceptible individuals come into contact with these areas. Military operations have often been severely affected by ST. During World War II, ST left a trail of sick soldiers in all the areas where allied soldiers were sent to contest the advances of the Japanese armies. In India, Burma, along the old Burma Road, and in the Philippines, there were 6861 cases in the American Army, 6730 among British and Indian troops, 3188 among the Australians, 613 in the US Navy, 176 cases among

**Fig. 7.12** Approximate geographic distribution of scrub typhus

Merrill's Marauders, and 349 cases among the Chinese [34]. In some areas, more than one out of every four men with the disease died. Casualties were so high that the Office of the Surgeon General prepared and sent out posters to combat areas detailing salient points of information about the dangers of ST and methods for prevention (Fig. 7.13).

## 7.4.2 Clinical and Laboratory Findings

The bite site is usually unremarkable at first, but after an incubation period of about 10 d, a papule may develop, eventually enlarging and undergoing central necrosis to form an eschar. An eschar (*see* Chap. 4, Fig. 4.13) is found in 48–82% of ST patients and is virtually pathognomonic when seen by a physician experienced in diagnosing ST [35, 36]. The acute febrile onset is characterized by headache, profuse sweating, conjunctival injection, and lymphadenopathy. There may be an accompanying maculopapular rash, which first appears on the trunk, later extending to the extremities. Photophobia, bronchitis, and cough are also frequently reported. Untreated, the disease may sometimes progress to deafness, anuria, pulmonary edema, or cardiac failure. Diagnosis is usually made based on history, clinical presentation, and serological tests, such as enzyme immunoassay (EIA) or indirect fluorescent antibody (IFA). A dipstick assay has been developed (Dip-Sticks), which is easy to perform and gives results in about 1 h [37]. PCR assays may also be used to diagnose ST. Some labs have the capability to isolate the infectious agent by inoculating the patient's blood into mice.

**Fig. 7.13** An educational poster used to inform soldiers on the dangers of mite infestations and on methods of personal protection. (US Army Medical Museum)

### 7.4.3  Ecology of ST and Its Vectors

ST is transmitted among wild rats by the larval stage (not a worm; first stage mites are called larvae) of trombiculid mites (Fig. 7.14). Larvae of the vector species—which are mostly in the genus *Leptotrombidium*—infest rodents and insectivores, and the distribution of the mites is dependent on the home ranges of the hosts (Table 7.2). These home ranges do not usually overlap; mite colonies therefore tend to be isolated from each other and occur as "mite islands." The rickettsiae are transstadially transmitted through nonparasitic nymphal and adult mite stages, which are predatory on soil arthropods and transovarially through the eggs to parasitic larvae of the next generation.

### 7.4.4  Treatment, Prevention, and Control

Area-wide mite control programs, using pesticides to spray ground and vegetation in camps and other rural settings, have limited success in controlling the vector mites. In addition, spraying pant legs and socks with permethrin-based

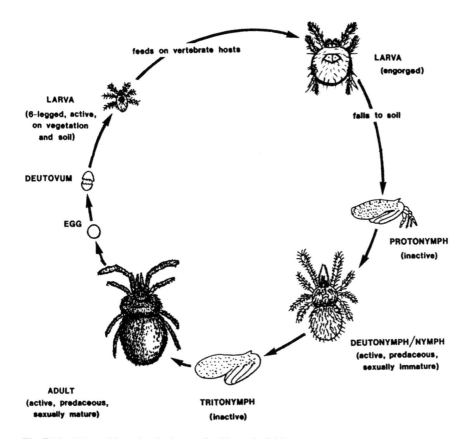

**Fig. 7.14** Chigger life cycle. (Redrawn after Varma Ref. 33)

aerosols (identical to tick repellent products; *see* Appendix 3) is an effective way to prevent contact with infected mites. Treatment of ST is with doxycycline or perhaps chloramphenicol [5, 35, 36]. However, there are areas in northern Thailand where chloramphenicol-resistant and doxycycline-resistant strains of *O. tsutsugamushi* occur [38].

## 7.5 Louse-Borne Infections

### 7.5.1 General and Medical Importance of Body Lice

The body louse, *Pediculus humanus humanus*, is a blood-feeding ectoparasite of humans (Fig. 7.15). Body and head lice look almost identical (and are most likely one species [39]), but head lice remain more or less on the scalp and body lice on the body or in clothing. Body lice are relatively rare among affluent

**Table 7.2** Some major trombiculid species that transmit the agent of ST

| Vector species | Approximate distribution | Remarks |
| --- | --- | --- |
| *Leptotrombidium deliense* | Almost entire area where scrub typhus occurs | Overall, principal vector |
| *L. akamushi* | Japan | Chief vector in Japan |
| *L. fletcheri* | Malaysia, New Guinea, Philippines | |
| *L. arenicola* | Malaysia found on sandy beaches | |
| *L. pallidum* | Parts of Japan | |
| *L. scutellaris* | Parts of Japan | |
| *L. pavlovsky* | Far eastern parts of Russia | |

**Fig. 7.15** The human body louse, vector of epidemic typhus, trench fever, and relapsing fever organisms. (Photo copyright 2005 by Jerome Goddard, PhD)

members of industrial nations, yet they can become severe under crowded and unsanitary conditions, such as war or natural disasters. Body lice may transmit the agent of *epidemic typhus*, and there have been devastating epidemics of the disease in the past. Typhus is still endemic in poorly developed countries, mostly in Africa, but occasional cases have been identified in Asia, South America, and even the United States [39, 40]. Besides louse-borne typhus, body lice transmit the agents of *trench fever* and *louse-borne relapsing fever*. Trench fever is still widespread in parts of Europe, Asia, Africa, and the Americas but often in an asymptomatic form [39–42]. Louse-borne relapsing fever may occur in many places in Europe and Asia but particularly in the Horn of Africa. Aside from the possibility of disease transmission, body lice may cause severe skin irritation. The usual clinical presentation is pyoderma in covered areas. Characteristically, some swelling and red papules develop at each bite site. There are intermittent episodes of mild to severe itch associated with the bites. Compounding this, some individuals become sensitized to antigens injected during louse biting, leading to generalized allergic reactions. Subsequent excoriation of the skin by the infested individual may lead to impetigo or eczema. Sometimes long-standing infestations lead to a brownish-bronze pigmentation of the skin, especially in the groin, axilla, and upper thigh regions.

## 7.5.2   Epidemic Typhus

Louse-borne (epidemic) typhus, caused by the rickettsial organism, *Rickettsia prowazekii*, is characterized by high fever for about 2 wk. accompanied by head-ache, chills, prostration, bronchial disturbance, and mental confusion [43]. In addition, a macular eruption often appears on the fifth to sixth day, initially on the upper trunk, followed by spread to the entire body, but usually not to the face, palms, or soles [5]. The case fatality rate increases with age and varies from 10% to 40%. Epidemic typhus may occur in many areas of the world (Fig. 7.16) and is associated with poverty, wars, and natural disasters because of poor hygiene, crowding, and extended wearing of the same clothing (a factor favorable to lice development). Interestingly, there was a case of *R. prowazekii* infection reported from south Texas in 2001 [44]. At times, the effects of typhus have been staggering. During World War I, Russia lost 2–3 million citizens to typhus [12]. In World War II, a large out-break threatened virtually to wipe out Naples, Italy, in September of 1943. Under the crowded, unsanitary conditions of Naples at that time, the death rate reached as high as 81% [12]. Outbreaks still occur today; there was a major outbreak of typhus in Burundi in the 1990s because of the civil war [45].

Body lice become infected by feeding on the blood of a person acutely ill with the disease, that is, having a high rickettsemia. Infected lice subsequently excrete

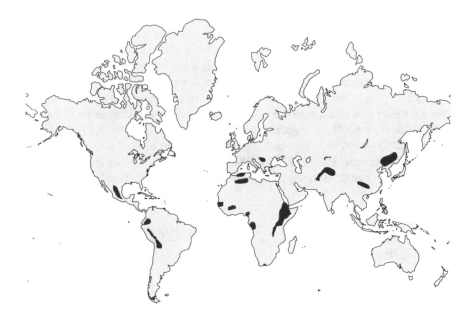

**Fig. 7.16** Approximate geographic distribution of louse-borne epidemic typhus

rickettsial organisms in their feces while feeding. Humans become infected by rubbing or scratching feces or crushed lice into superficial abrasions on the skin.

### 7.5.3   Trench Fever

Trench fever, caused by *Bartonella (=Rochalimaea) quintana*, is characterized by fever, rash, bone pain (especially the shins), and splenomegaly and ranges in severity from a mild flu-like illness to a more severe, relapsing disease [46]. It is generally nonfatal. The organism was first identified as a human pathogen when it caused at least 1 million cases among troops in Europe during World War I [47]. Recently, the disease has reemerged among homeless persons in North America and Europe and has also been found to cause bacillary angiomatosis, endocarditis, and bacteremia in HIV-infected persons. As in the case of typhus, epidemics depend on heavy body lice infestations in a susceptible human population living in low socioeconomic conditions.

### 7.5.4   Louse-Borne Relapsing Fever (LBRF)

LBRF, caused by the spirochete, *Borrelia recurrentis*, is a systemic spirochetal disease characterized by periods of fever lasting 2–9 d alternating with afebrile periods of 2–4 d. Symptoms include high fever, headache, prostration, myalgias, and sometimes gastrointestinal manifestations. LBRF is very similar (if not the same thing) to tick-borne relapsing fever (*see* Sect. 4.9. in Chap. 4). Transmission of the spirochetes is not by bite, but instead, they are introduced in several ways: at the bite site (by crushing the lice), the skin of the crushing fingers, the conjunctivae when people rub their eyes, or through mucous membranes of the mouth (people sometimes bite lice to kill them) [48]. The disease is theoretically cosmopolitan but is particularly prevalent in the Ethiopian region (Fig. 7.17). Between 2000 and 5000 cases were reported annually from Ethiopia and a smaller number from Sudan between 1967 and 1971 [12]. Some authors estimate that millions of cases of LBRF occurred during the two world wars of the twentieth century [48].

### 7.5.5   Treatment, Prevention, and Control of Louse-Borne Diseases

Doxycycline and chloramphenicol are effective against typhus and probably effective against trench fever, although gentamicin is sometimes used along with doxycycline for trench fever [5, 46]. Doxycycline and erythromycin are effective treatments for relapsing fever [5, 48]. Concurrent with treatment, patients need to be disinfected (or disinfested) by the use of insecticidal dusts or sprays and

**Fig. 7.17**  Approximate geographic distribution of louse-borne relapsing fever

**Fig. 7.18**  Army soldier spraying DDT into clothing of a refugee to kill body lice. (US Army photo)

laundering of clothing and bed covers. Sometimes, community or military delousing campaigns include hand or power blowers to apply an effective residual insecticide powder to people and their clothing such as was done during World War II (Fig. 7.18).

# References

1. Cantey PT, Stramer SL, Townsend RL, Kamel H, Ofafa K, Todd CW, et al. The United States *Trypanosoma cruzi* infection study: evidence for vector-borne transmission of the parasite that causes Chagas' disease among United States blood donors. Transfusion. 2012;52:1922–30.
2. Gunter SM, Murray KO, Gorchakov R, Beddard R, Rossmann SN, Montgomery SP, et al. Likely autochthonous transmission of *Trypanosoma cruzi* to humans, south central Texas, USA. Emerg Infect Dis. 2017;23:500–3.
3. Waleckx E, Suarez J, Richrads B, Dorn PL. *Triatoma sanguisuga* blood meals and potential for Chagas' disease, Louisiana, USA. Emerg Infect Dis. 2014;20:2141–3.
4. Bern C. Chagas' disease. N Engl J Med. 2015;373:456–66.
5. Heymann DL, editor. Control of communicable diseases manual. 20th ed. Washington, DC: American Public Health Association; 2015.
6. Spira AM. Trypanosomiasis Part 2: Chagas' disease. Inf Med. 2006;23:219–21.
7. Kirchhof LV. American trypanosomiasis (Chagas' disease). In: Guerrant RL, Walker DH, Weller PF, editors. Tropical infectious diseases: principles, pathogens, and practice. 3rd ed. New York: Saunders Elsevier; 2011. p. 689–96.
8. Markell E, Voge M, John D. Medical parasitology. 7th ed. Philadelphia: W.B. Saunders; 1992.
9. Cardoni RL. Inflammatory response to acute *Trypanosoma cruzi* infection. Medicina. 1997;57:227–34.
10. Ribeiro AL, Rocha MO. Indeterminate form of Chagas disease: considerations about diagnosis and prognosis. Rev Soc Bras Med Trop. 1998;31:301–14.
11. Carriazo CS, Sembaj A, Aguerri AM, Requena JM, Alonso C, Bua J, et al. Polymerase chain reaction procedure to detect *Trypanosoma cruzi* in blood samples from chronic chagasic patients. Diagn Microbiol Infect Dis. 1998;30:183–6.
12. Harwood RF, James MT. Entomology in human and animal health, 7th Edition. 7th ed. New York: Macmillan; 1979. 548 p
13. Guzman-Bracho C, Lahuerta S, Velasco-Castrejon O. Chagas disease: first congenital case report. Arch Med Res. 1998;29:195–6.
14. Rossi AJ, Rossi A, Marin-Neto JA. Chagas' disease. Lancet. 2010;375:1388–402.
15. Schettino PMS, Arteaga IDH, Berrueta TU. Chagas disease in Mexico. Parasitol Today. 1988;4:348–52.
16. Zulantay I, Apt W, Rodriguez J, Venegas J, Sanchez G. Serologic evaluation of treatment of chronic Chagas disease with allopurinol and itraconazole. Rev Med Chil. 1998;126:265–70.
17. Maguire JH. Treatment of Chagas' disease – time is running out. N Engl J Med. 2015;373:1369–70.
18. Dorn PL, Buekens P, Hanford E. Whac-amole: future trends in Chagas' transmission and the importance of a global perspective on disease control. Future Microbiol. 2007;2:365–7.
19. Gubler DJ. Resurgent vector-borne diseases as a global health problem. Emerg Infect Dis. 1998;4:442–9.
20. Buscher P, Cecchi G, Jamonneau V, Priotto G. Human African trypanosomiasis. Lancet. 2017;390(10110):2397–409.
21. Pepin J, Donelson JE. African trypanosomiasis. In: Guerrant RL, Walker DH, Weller PF, editors. Tropical infectious diseases. 3rd ed. Philadelphia: Elsevier Saunders; 2011. p. 682–8.
22. Dumas M, Bouteille B. Current status of trypanosomiasis. Med Trop (Mars). 1997; 57(3 Suppl):65–9.
23. Enserink M. Welcome to Ethiopia's fly factory. Science (News Focus). 2007;317:310–3.
24. WHO. Progress report on the elimination of human onchocerciasis, 2016–2017. Wkly Epidemiol Rec. 2017;92(45):681–94.
25. Abiose A. Onchocercal eye disease and the impact of Mectizan treatment. Ann Trop Med Parasitol. 1998;92(Suppl 1):11–22.
26. Molyneux DH. Vector-borne parasitic diseases – an overview of recent changes. Int J Parasitol. 1998;28:927–34.

27. Hoerauf AM. Onchocerciasis. In: Guerrant RL, Walker DH, Weller PF, editors. Tropical infectious diseases. 3rd ed. Philadelphia: Elsevier Saunders; 2011. p. 741–9.

28. Crosskey RW. Blackflies. In: Lane RP, Crosskey RW, editors. Medical insects and arachnids. London: Chapman and Hall; 1996. p. 241–87.

29. Choudhary IA, Choudhary S. Resistant pruritis and rash in an immigrant. Inf Med. 2005;22:187–9.

30. Weitzel T, Dittrich S, Lopez J, Phuklia W, Martinez-Valdebenito C, Velasquez K, et al. Endemic scrub typhus in South America. New Engl J Med. 2016;375:954–61.

31. Walker DH. Scrub typhus – scientific neglect, ever-widening impact. New Engl J Med. 2016;375:913–5.

32. Kelly DJ, Fuerst PA, Ching WM, Richards AL. Scrub typhus: the geographic distribution of phenotypic and genotypic variants of *Orientia tsutsugamushi*. Clin Infect Dis. 2009; 48(Suppl. 3):S203–S30.

33. Varma MGR. Ticks and mites. In: Lane RP, Crosskey RW, Eds., editors. Medical insects and arachnids. London: Chapman and Hall; 1993. p. chap. 18.

34. Cushing E. History of entomology in World War II. Washington, DC: Smithsonian Institution; 1957. 117 p

35. Gormley TS. A diagnosis of scrub typhus. Navy Medicine. 1996;(November–December): 20–2.

36. Watt G, Walker DH. Scrub typhus. In: Guerrant RL, Walker DH, Weller PF, editors. Tropical infectious diseases: principles, pathogens, and practice. 1. 2nd ed. Philadelphia: Churchill Livingstone; 2006. p. 557–63.

37. Pradutkanchana J, Silpapojakul K, Paxton H, Pradutkanchana S, Kelly DJ, Strickman D. Comparative evaluation of four diagnostic tests for scrub typhus in Thailand. Trans Royal Soc Trop Med Hyg. 1997;91:425–8.

38. Watt G, Chouriyagune C, Ruangweerayud R, Watcharapichat P, Phulsuksombati D, Jongsakul K, et al. Scrub typhus infections poorly responsive to antibiotics in northern Thailand. Lancet. 1996;348:86–9.

39. Bonilla DL, Durden L, Eremeeva ME, Dasch GA. The biology and taxonomy of head and body lice: implications for louse-borne disease prevention. PLoS Pathog. 2013;9:1–5.

40. Murray KO, Evert N, Mayes B, Fonken E, Erickson T, Garcia MN, et al. Typhus group rickettsiosis, Texas, USA, 2003-2013. Emerg Infect Dis. 2017;23(4):645–8.

41. Dehio C, Maguina C, Walker DH. Bartonelloses. In: Guerrant RL, Walker DH, Weller PF, editors. Tropical infectious diseases: principles, pathogens, and practice. 3rd ed. New York: Saunders Elsevier; 2011. p. 265–71.

42. Faccini-Martinez AA, Marquez AC, Bravo-Estupinan DM, Calixto O, Botero-Garcia CA, Hidalgo M, et al. *Bartonella quintana* and typhus group rickettsiae exposure among homeless persons, Bogata, Colombia. Emerg Infect Dis. 2017;23:1876–9.

43. Walker DH, Raoult D. Typhus group rickettsioses. In: Guerrant RL, Walker DH, Weller PF, editors. Tropical infectious diseases: principles, pathogens, and practice. 3rd ed. New York: Saunders Elsevier; 2011. p. 329–34.

44. Massung RF, Davis LE, Slater K, McKechnie DB, Puerzer M. Epidemic typhus meningitis in the southwestern United States. Clin Infect Dis. 2001;32(6):979–82.

45. Raoult D, Ndihokubwayo JB, Tissot-Dupont H, Roux V, Faugere B, Abegbinni R, et al. Outbreak of epidemic typhus associated with trench fever in Burundi. Lancet. 1998;352:353–8.

46. Walker DH, Maguina C, Minnick M. Bartonelloses. In: Guerrant RL, Walker DH, Weller PF, editors. Tropical infectious diseases: principles, pathogens, and practice, vol. 1. Philadelphia: Churchill Livingstone; 2006. p. 454–62.

47. Jackson LA, Spach DH. Emergence of *Bartonella quintana* infection among homeless persons. Emerg Infect Dis. 1996;2:141–4.

48. Barbour A. Relapsing fever and other *Borrelia* diseases. In: Guerrant RL, Walker DH, Weller PF, editors. Tropical infectious diseases: principles, pathogens, and practice. 3rd ed. New York: Saunders Elsevier; 2011. p. 295–302.

# Part III
# Other Arthropod-Caused or -Related Problems

# Chapter 8
# Health Effects of Bed Bugs

## 8.1 Introduction and Bed Bug Biology

The common bed bug, *Cimex lectularius*, has been associated with humans for thousands of years. The word *Cimex* is derived from the Roman designation for bug and *lectularius* from the Latin name for couch or bed [1]. Bed bugs are common in the developing world and especially in areas of extreme poverty and crowding. The bloodsucking parasites had nearly disappeared in developed countries until fairly recently when studies showed a dramatic spread since the 1980s [2, 3]. The parasites have been reported as increasingly common inside US hotel rooms, dorms, and apartments [4–7] and even in "odd" places such as emergency rooms, subways, and other forms of transportation [8–10]. Recent studies have highlighted the problem of bed bugs being found in hospitals and the high cost of decontaminating infested rooms [9, 11]. Figure 8.1 shows a bed bug infestation of a mattress and box springs.

Bed bugs are cosmopolitan in distribution, mostly found in temperate regions worldwide [12]. Another human-biting bed bug species, *Cimex hemipterus*, is also widespread but is found mostly in the tropics. Several other bed bug species are found on bats, but they do not usually bite people [13]. Adult bed bugs are approximately 5 mm long, oval, and flattened. They somewhat resemble unfed ticks or small cockroaches. Adults are reddish brown (chestnut) (Fig. 8.2); immature bugs resemble adults but are much smaller and may be yellowish white in color (Fig. 8.3). Bed bugs have a pyramid-shaped head with prominent compound eyes, slender antennae, and a long proboscis tucked backward underneath the head and thorax (Fig. 8.4). The prothorax (dorsal side, first thoracic segment) bears rounded, wing-like lateral horns on each side.

Bed bugs possess stink glands and emit an odor. Homes heavily infested with the bugs may have this distinct odor. Bed bugs generally feed at night, hiding in crevices during the day. Hiding places include seams in mattresses, crevices in box springs, spaces under baseboards or loose wallpaper, and behind boards and debris (Fig. 8.5). There are five nymphal stages that must be passed before development to

© Springer International Publishing AG, part of Springer Nature 2018
J. Goddard, *Infectious Diseases and Arthropods*, Infectious Disease,
https://doi.org/10.1007/978-3-319-75874-9_8

**Fig. 8.1** Bed bugs and
fecal spots on box springs.
(Photo courtesy Armed
Forces Pest Management
Board, taken by Harold
Harlan)

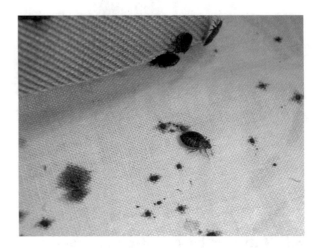

**Fig. 8.2** Bed bugs before
and after feeding. (Photo
copyright 2008 by Jerome
Goddard, Ph.D.)

adulthood. Once an adult, the bed bug has a life-span of 6–12 mo. At each nymphal
stage, the bed bug must take a blood meal in order to complete development and
molt to the next stage. The bugs take about 5–10 min to ingest a full blood meal.
Bed bugs can survive long periods without feeding, and when their preferred human
hosts are absent, they may take a blood meal from any warm-blooded animal.

## 8.2   Reactions to Bed Bug Bites

As in other members of the insect order Hemiptera, bed bugs have piercing/suck-
ing mouthparts. Accordingly, bites from the bugs may produce welts and local
inflammation, probably from allergic reactions to saliva injected during feeding
(Fig. 8.6) [14–18]. On the other hand, in many persons, the bite is undetectable and
produces no lesion [17, 19]. Bed bugs bites may affect human health in a number

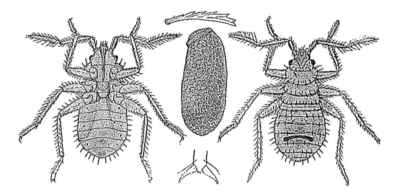

**Fig. 8.3** Bed bug life stages, showing nymphs and eggs. (U.S.D.A. Bulletin Number 4, Washington, D.C.)

**Fig. 8.4** Bed bug proboscis. (Photo copyright 2012 by Jerome Goddard, Ph.D.)

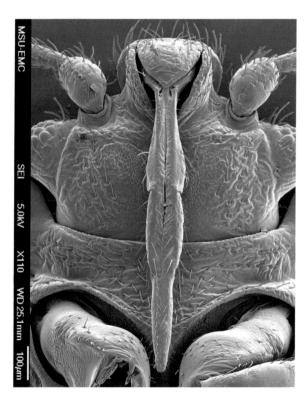

of ways (see textbox), but the principal medical complaint is emotional trauma from biting incidents and the itching and inflammation associated with their bites. There is limited evidence that bed bug attacks may cause psychological effects similar to posttraumatic stress disorder [20]. The most commonly reported cutaneous reactions are small, pruritic maculopapular, erythematous lesions at bed bug

**Fig. 8.5** Bed bugs on a piece of wood from an infested house. (Photo copyright 2012 by Jerome Goddard, Ph.D.)

**Fig. 8.6** Bed bug bites on ankle. (Photo courtesy Kristine T. Edwards, used with permission)

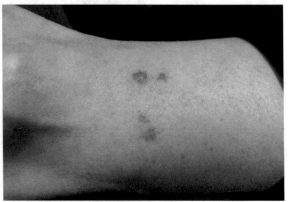

feeding sites, one per insect. These usually itch and, if not scratched extensively, resolve within 10 days [17, 21, 22]. The intensity of these bite reactions may increase in individuals who experience repeated bites [23–25]. Some people experience more complex, even serious, cutaneous reactions. Reports of these have included pruritic wheals (local urticaria) around a central punctum, papular urticaria, and diffuse urticaria at bite sites usually noted on arising in the morning [26–30]. Bullous lesions may occur upon subsequent biting events days later [8, 14, 31–33]. The timing of cutaneous reactions to bed bugs may change (shorten) with subsequent exposures which appears to reflect host immunological responses to bed bug salivary proteins [14, 15, 17, 34–37]. Usinger [12] fed a colony of bed bugs on himself weekly for 7 years and noted that his reactions progressed from delayed to immediate, with no evidence of desensitization. Reinhardt et al. [38] also reported that with repeated exposure, the latency between bite and skin reaction decreased from 10 days to a few seconds. Interestingly, people who experience subsequent bed bug biting sometimes have old lesions that become red and itchy again upon new biting anywhere on the body. This "relighting-up" phenomenon at sites of previous lesions has been anecdotally reported previously [18, 39] but is poorly understood. It presumably results from antigens residing at the bite site for an extended period of time and from these antigens responding to inflammatory mediators circulating to new bites.

Although rare, there are a few reports of systemic reactions from bed bug bites, including asthma, generalized urticaria, and anaphylaxis [32, 40–42]. One patient fed 40–50 bed bugs on himself, and after 8 min, he developed itch, swelling of the face, lethargy, profuse sweating, and widespread urticaria on the torso and limbs [43]. Another patient awakened during the night at a hotel with severe itching and urticaria on his arm and neck, and bed bugs were found in his room [44]. He developed angioedema and hypotension, was hospitalized, and had transient anterolateral ischemia on electrocardiogram. Eight months later, after an experimental bed bug bite, he developed a wheal at the bite site and generalized itching that required epinephrine administration to resolve his symptoms.

**Five Ways Bed Bugs May Affect Human Health**
1. Emotional/psychological effects
2. Nuisance biting (non-allergic bite reactions)
3. Anemia
4. Systemic allergic reactions (anaphylactoid reactions)
5. Possible disease transmission

## 8.3   Bed Bugs and Disease Transmission

The possibility of transmission of human disease agents by bed bugs is controversial. Since the insects repeatedly suck blood from humans and live a relatively long time, conceivably they might ingest a pathogen and later transmit it. Burton [45] reported that bed bugs have been suspected in the transmission of 41 human diseases; however, finding a bloodsucking insect infected with a pathogen does not mean that it is a competent vector of that agent or even a vector at all [46]. Sometimes after imbibing an infectious blood meal, pathogens in the blood may survive a short period of time, or bits and pieces of them may be detectable. In our laboratory, we have fed bed bugs on chicken blood containing the bacteria, *Rickettsia parkeri*, and, after 2 weeks, found *R. parkeri*-like organisms in the bed bugs' salivary glands [47] (Fig. 8.7). But this fact does not prove transmissibility of the agent.

There have been studies of possible HIV transmission by bed bugs. Webb and colleagues [48] found that HIV could be detected in bed bugs up to 8 d after exposure to highly concentrated virus in blood meals, but no viral replication was

**Fig. 8.7** Artificial feeding of bed bugs on chicken blood infected with the bacteria, *Rickettsia parkeri* (left), and *Rickettsia*-like organisms seen in bed bug salivary glands 2 weeks later. (Photo copyright 2012 by Jerome Goddard, Ph.D.)

observed nor was any virus detected in bed bug feces. In addition, by using an artificial system of feeding bed bugs through membranes, the authors could not demonstrate mechanical transmission of HIV.

One of the best candidates for transmission by bed bugs is hepatitis B virus (HBV). Pools (groups) of bed bugs collected from huts in Northern Transvaal, South Africa—an area with high rates of human HBV seropositivity—tested positive for hepatitis B surface antigen (HBsAG) [49]. In addition, HBsAG has been shown to persist in bed bugs for at least 7.5 wk. after experimental feeding [50, 51]. However, Jupp and McElligott [50] found no biologic multiplication of HBV in bed bugs. Polymerase chain reaction (PCR) assays have detected HBV DNA in bed bugs and their excrement up to 6 wk. after feeding on an infectious meal [52, 53]. Another study suggested that bed bug feces might be considered a source of mechanical transmission of HBV infection under some circumstances [54]. However, finding HBV surface antigen or PCR amplicons (amplified pieces of DNA matching primers used in the test) in feces is no indication of viable virus. These could be "digested" pieces/parts and not live virus. Further, a 2-year intervention in Gambia wherein insecticides were sprayed extensively inside human dwellings reduced exposure to bed bugs but had no effect on HBV infection [55].

Whether HBV infectivity survives the bed bug digestive process is unknown. A transmission experiment with chimpanzees helped resolve this issue, although the sample size was small (only three animals) [56]. In that study, bed bugs were fed HBV-infected blood through a membrane. Ten to 13 days later, subsamples of the bugs were tested for infectivity; 53–83% were found to be infected. Then, ~200 of the infected bugs took meals from the three chimpanzees. No infections or seroconversions resulted. To confirm infectivity of the inoculum, the researchers then injected the same three animals with a portion of the original blood used to infect the bed bugs. HBV infections followed quickly in all three chimpanzees.

Recent reports have raised the possibility of transmission of MRSA, Chagas' disease, and certain viruses by bed bugs [57–59], but further work is needed to elucidate those possibilities. Whether or not bed bugs transmit human disease agents remains a point of contention although statements in most mainstream scientific papers on the subject say they do not [2, 16]. Attorneys representing plaintiffs bitten by the bugs in hotels or apartments often firmly state that the risk is real and warrants compensation. However, until further evidence proves otherwise, I think the best summary of current data goes something like this: "Currently, there is little evidence of disease transmission by bed bugs."

# References

1. Butler E. Our household insects. London: Longmans, Green, and Co.; 1893. 344 p.
2. Reinhardt K, Siva-Jothy MT. Biology of the bed bugs. Annu Rev Entomol. 2007;52:351–74.
3. Miller D. Bed bugs (Hemiptera: Cimicidae). In: Capinera JL, editor. Encyclopedia of entomology. New York: Springer International; 2008. p. 405–17.
4. CHPPM. Bed bugs. U. S. Army Center for Health Promotion and Preventive Medicine (USACHPPM), Aberdeen Proving Ground, Maryland, Facts Sheet Number 18-029-0207; 2005.
5. Hwang SW, Svoboda TJ, De Jong IJ, Kabasele KJ, Gogosis E. Bed bug infestations in an urban environment. Emerg Infect Dis. 2005;11:533–8.
6. Potter MF. The perfect storm: an extension view on bed bugs. Am Entomol. 2006;52:102–4.
7. Potter MF, Rosenberg B, Henriksen M. Bugs without borders: defining the global bed bug resurgence. Pest World, Sept/Oct. 2010:8–20.
8. Kinnear J. Epidemic of bullous erythema on legs due to bed-bug. Lancet. 1948;255:55.
9. Sheele JM, Gaines S, Maurer N, Coppolino K, Li JS, Pound A, et al. A survey of patients with bed bugs in the emergency department. Am J Emerg Med. 2017;3(16):30998–9.
10. UPI. Bedbugs seen at New York subway stations. United Press International, May 8, http://www.upi.com/NewsTrack/Quirks/2008/05/08/bedbugs_seen_at_new_york_subway_stations/5081/; 2008.
11. Totten V, Charbonneau H, Hoch W, Sheele JM. The cost of decontaminating an ED after finding a bed bug: results from a single academic medical center. J Emerg Med. 2016;34:649.
12. Usinger RL. Monograph of Cimicidae. College Park, MD: Entomological Society of America, Thomas Say Foundation; 1966. 334 p.
13. Goddard J, Baker GT, Ferrari F, Ferrari C. Bed bugs and bat bugs: a confusing issue. Outlook Pest Manag. 2012;23:125–7.
14. Cooper DL. Can bedbug bites cause bullous erythema? J Am Med Assoc. 1948;138:1206.
15. Elston DM, Stockwell S. What's eating you? Bed bugs. Cutis. 2000;65:262–4.
16. Goddard J, de Shazo RD. Bed bugs (Cimex lectularius) and clinical consequences of their bites. J Am Med Assoc. 2009;301:1358–66.
17. Ryckman RE. Dermatological reactions to the bites of four species of triatominae (hemiptera: reduviidae) and Cimex lectularius L. (hemiptera: cimicidae). Bull Soc Vector Ecol. 1985;10:122–5.
18. Goddard J, Edwards KT, de Shazo RD. Observations on development of cutaneous lesions from bites by the common bed bug, Cimex lectularius L. Midsouth Entomol. 2011;4:49–52.
19. Goddard J, de Shazo RD. Multiple feeding by the common bed bug, Cimex lectularius L., without sensitization. Midsouth Entomol. 2009;2:90–2.
20. Goddard J, de Shazo RD. Psychological effects of bed bug attacks (Cimex lectularius L.). Am J Med. 2012;125(1):101–3.
21. Kemper H. Die Bettwanze und ihre bekampfung. Schriften uber Hygienische Zoologie, Z Kleintierk Pelztierk. 1936;12:1–107.

22. Masetti M, Bruschi F. Bedbug infestations recorded in Central Italy. Parasitol Int. 2007;56(1):81–3.
23. Liebold K, Schliemann-Willers S, Wollina U. Disseminated bullous eruption with systemic reaction caused by *Cimex lectularius*. J Eur Acad Dermatol Venereol. 2003;17:461–3.
24. Bartley JD, Harlan HJ. Bed bug infestation: its control and management. Mil Med. 1974;139:884–6.
25. Cestari TF, Martignago BF. Scabies, pediculosis, and stinkbugs: uncommon presentations. Clinics Dermatol. 2005;23:545–54.
26. Alexander JO. Arthropods and human skin. Berlin: Springer; 1984. 422 p.
27. Brasch J, Schwarz T. 26-year-old male with urticarial papules. J Dtsch Dermatol Ges. 2006;4(12):1077–9.
28. Gbakima AA, Terry BC, Kanja F, Kortequee S, Dukuley I, Sahr F. High prevalence of bedbugs *Cimex hemipterus* and *Cimex lectularis* in camps for internally displaced persons in Freetown, Sierra Leone: a pilot humanitarian investigation. West Afr J Med. 2002;21(4):268–71.
29. Honig PJ. Arthropod bites, stings, and infestations: their prevention and treatment. Pediatr Dermatol. 1986;3:189–97.
30. Rook AJ. Papular urticaria. Ped Clin NA. 1961;8:817–20.
31. Hamburger F, Dietrich A. Lichen urticatus exogenes. Acta Paediatr. 1937;22:420.
32. Kemper H. Beobachtungen ueber den Stech-und Saugakt der Bettwanze und seine Wirkung auf die menschliche Haut. Zeitschr f Desinfekt. 1929;21:61–5.
33. Patton WS, Evans A. Insects, ticks, and venomous animals of medical and veterinary importance, part I. Croydon, U.K.: H.R. Grubb Ltd.; 1929.
34. Churchill TP. Urticaria due to bed bug bites. J Am Med Assoc. 1930;95:1975–6.
35. Hect O. Die hautreaktionen auf insektenstiche als allergische erscheinungen. Zool Anz. 1930;87:94, 145, 231.
36. Ryckman RE. Host reactions to bug bites (hemiptera, homoptera): a literature review and annotated bibliography, Part I. California Vect Views. 1979;26:1–24.
37. Ryckman RE, Bently DG. Host reactions to bug bites: a literature review and annotated bibliography, Part II. California Vect Views. 1979;26:25–49.
38. Reinhardt K, Kempke RA, Naylor RA, Siva-Jothy MT. Sensitivity to bites by the bedbug, *Cimex lectularius*. Med and Vet Entomol. 2009;23:163–6.
39. McKiel JA, West AS. Nature and causation of insect bite reactions. Ped Clin NA. 1961;8:795–814.
40. Bircher AJ. Systemic immediate allergic reactions to arthropod stings and bites. Dermatology. 2005;210(2):119–27.
41. Jimenez-Diaz C, Cuenca BS. Asthma produced by susceptibility to unusual allergens. J Allergy. 1935;6:397–403.
42. Sternberg L. A case for asthma caused by the *Cimex lectularius*. Med J Rec. 1929;129:622.
43. Minocha R, Wang C, Dang K, Webb CE, Fernández-Peñas P, Doggett SL. Systemic and erythrodermic reactions following repeated exposure to bites from the common bed bug *Cimex lectularius* (Hemiptera: Cimicidae). Austral Entomol. 2016. https://doi.org/10.1111/aen.12250.
44. Parsons DJ. Bed bug bite anaphylaxis misinterpreted as coronary occlusion. Ohio State Med J. 1955;51:669.
45. Burton GJ. Bed bugs in relation to transmission of human diseases. Public Health Rep. 1963;78:513–24.
46. Goddard J. Mosquito vector competence and West Nile virus transmission. Inf Med. 2002;19:542–3.
47. Goddard J, Varela-Stokes A, Smith W, Edwards KT. Artificial infection of the bed bug with *Rickettsia parkeri*. J Med Entomol. 2012;49:922–6.
48. Webb PA, Happ CM, Maupin GO, Johnson B, Ou C-Y, Monath TP. Potential for insect transmission of HIV: experimental exposure of *Cimex hemipterus* and *Toxorhynchites amboinensis* to human immunodeficiency virus. J Infect Dis. 1989;160:970–7.

49. Jupp PG, Prozesky OW, McElligott SE, Van Wyk LA. Infection of the common bed bug with hepatitis B virus in South Africa. S Afr Med J. 1978;53:598–600.
50. Jupp PG, McElligott SE. Transmission experiments with hepatitis B surface antigen and the common bed bug. S Afr Med J. 1979;56:54–7.
51. Newkirk MM, Downe AER, Simon JB. Fate of ingested hepatitis B antigen in blood-sucking insects. Gastroenterol. 1975;69:982–7.
52. Blow JA, Turell MJ, Silverman AL, Walker E. Stercorarial and transtadial transmission of hepatitis B virus by common bed bugs. J Med Entomol. 2001;38:694–700.
53. Silverman AL, Qu LH, Blow J, Zitron IM, Gordon SC, Walker ED. Assessment of hepatitis B virus DNA and hepatitis C virus RNA in the common bedbug (*Cimex lectularius* L.) and kissing bug (*Rodnius prolixus*). Am J Gastroenterol. 2001;96(7):2194–8.
54. Ogston CW, London WT. Excretion of hepatitis B surface antigen by the bed bug. Trans Royal Soc Trop Med Hyg. 1980;74:823–5.
55. Mayans MV, Hall AJ, Inskip HM. Do bedbugs transmit hepatitis B? Lancet. 1994;344:125.
56. Jupp PG, Purcell RH, Phillips JM, Shapiro M, Gerin JL. Attempts to transmit hepatitis B virus to chimpanzees by arthropods. S Afr Med J. 1991;79:320–2.
57. Salazar R, Castillo-Neyra R, Tustin AW, Borrini-Mayori K, Naquira C, Levy MZ. Bed bugs (*Cimex lectularius*) as vectors of *Trypanosoma cruzi*. Am J Trop Med Hyg. 2015;92(2):331–5.
58. Adelman ZN, Miller DM, Myles KM. Bed bugs and infectious disease: a case for the arboviruses. PLoS Pathog. 2013;9(8):e1003462.
59. Lowe CF, Romney MG. Bedbugs as vectors for drug-resistant bacteria. Emerg Infect Dis. 2011;17(6):1132–4.

# Chapter 9
# Why Mosquitoes and Other Arthropods Cannot Transmit HIV

Human immunodeficiency virus (HIV), the etiologic agent of AIDS, is an enveloped, positive-stranded RNA retrovirus [1]. Because HIV is a blood-borne pathogen, concerns have been raised about its possible transmission by blood-feeding arthropods. This chapter explores that possibility (Note: much of this discussion comes from a review article by McHugh [2] with his kind permission). First, distinction should be made between mechanical and biological transmission of disease agents (*see* Chap. 2). Mechanical transmission occurs when arthropods physically carry pathogens from one place or host to another, while in biological transmission, there is either multiplication or development of the pathogen. For biological transmission, the virus must avoid digestion in the gut of the insect, recognize receptors on and penetrate the gut, replicate in insect tissue, recognize and penetrate the insect salivary glands, and escape into the lumen of the salivary duct. In one study by Webb and colleagues, HIV virus persisted for 8 d in bed bugs [3]. Another study by Humphrey-Smith and colleagues showed the virus to persist for 10 d in ticks [4] artificially fed meals with high levels of virus ($\geq 10^5$ tissue culture infective doses/mL [TCID/ mL]), but there was no evidence of viral replication. Intra-abdominal inoculation of bed bugs and intrathoracic inoculation of mosquitoes were used to bypass any gut barriers, but again the virus failed to multiply [3]. Likewise, in vitro culture of HIV with a number of arthropod cell lines indicated that HIV was incapable of replicating in these systems. Thus, biological transmission of HIV seems extremely improbable.

Mechanical transmission would mostly likely occur if the arthropod were interrupted while feeding and then quickly resumed feeding on a susceptible host. Transmission of HIV would be a function of the viremia in the infected host and the virus remaining on the mouthparts or regurgitated into the feeding wound. The blood meal residue on bed bug mouthparts was estimated to be $7 \times 10^{-5}$ mL, but 50 bed bugs, interrupted while feeding on blood containing $1.3 \times 10^5$ TCID/mL HIV, failed to contaminate the uninfected blood on which they finished feeding or the mouse skin membrane through which they refed [3].

© Springer International Publishing AG, part of Springer Nature 2018
J. Goddard, *Infectious Diseases and Arthropods*, Infectious Disease,
https://doi.org/10.1007/978-3-319-75874-9_9

Within minutes of being fed blood with $5 \times 10^4$ TCID of HIV, stable flies regurgitated 0.2 FL of fluid containing an estimated 10 TCID [5]. The minimum infective dose for humans contaminated in this manner is unknown, but under conditions such as those in some tropical countries where there are large populations of biting insects and a high prevalence of HIV infection, transfer might be theoretically possible, if highly unlikely. In these countries, however, other modes of transmission are overwhelmingly important, and although of grave importance to the extremely rare individual who might contract HIV through an arthropod bite, arthropods are of no significance to the ecology of this virus.

An epidemiologic survey of Belle Glade, a south Florida community believed to have a number of HIV infections in individuals with no risk factors, provided no evidence of HIV transmission by insects [6]. Interviews with surviving patients with the infections revealed that all but a few had engaged in the traditional risk behavior (e.g., drug use and unprotected sex). A serosurvey for exposure to mosquito-borne viruses demonstrated no significant association between mosquito contact and HIV status nor were repellent use, time outdoors, or other factors associated with exposure to mosquitoes related to risk of HIV infection. A serosurvey for HIV antibodies detected no positive individuals between 2 and 10 yr of age or 60 and older. No clusters of cases occurred in houses without other risk factors. There was thus no evidence of insect-borne HIV.

# References

1. Reynolds SJ, Bessong PO, Quinn TC. Human retroviral infections in the tropics. In: Guerrant RL, Walker DH, Weller PF, editors. Tropical infectious diseases: principles, pathogens, and practice, vol. 1. 2nd ed. Philadelphia: Churchill Livingstone; 2006. p. 852–83.
2. McHugh CP. Arthropods: vectors of disease agents. Lab Med. 1994;25:429–37.
3. Webb PA, Happ CM, Maupin GO, Johnson B, Ou C-Y, Monath TP. Potential for insect transmission of HIV: experimental exposure of *Cimex hemipterus* and *Toxorhynchites amboinensis* to human immunodeficiency virus. J Infect Dis. 1989;160:970–7.
4. Humphery-Smith I, Donker G, Turzo A, Chastel C, Schmidt-Mayerova H. Evaluation of mechanical transmission of HIV by the African soft tick, *Ornithodoros moubata*. AIDS. 1993;7:341–7.
5. Brandner G, Kloft WJ, Schlager-Vollmer C, Platten E, Neumann-Opitz P. Preservation of HIV infectivity during uptake and regurgitation by the stable fly, *Stomoxys calcitrans*. AIDS-Forschung. 1992;5:253–6.
6. Castro KG, Lieb S, Jaffe HW. Transmission of HIV in belle glade, Florida: lessons for other communities in the United States. Science. 1988;239:193–7.

# Chapter 10
# Medical Conditions Caused by Arthropod Stings or Bites

## 10.1 Introduction and Medical Significance

Arthropods cause a wide variety of clinical conditions in humans, but especially skin lesions, because people are inevitably exposed to biting and stinging organisms in the urban and suburban environment [1–5]. Skin lesions resulting from arthropod exposure may arise via various pathologic pathways, such as direct damage to tissue, hypersensitivity reactions to venom or saliva, or infectious disease. Direct injury can occur from mouthparts or stingers piercing human skin [6] and/or blisters or stains resulting from exposure to arthropods [7, 8]. In some cases, proteins in venom or saliva may cause direct mast cell degranulation, leading to urticaria [9]. In addition, secondary infections may result from bacteria entering the skin via the bite/sting punctum. This is especially likely if the bite/sting site is scratched extensively. As discussed in Part II, many vector-borne infectious diseases can also produce skin lesions such as rash, ulcers, or eschar.

## 10.2 Pathogenesis

### 10.2.1 Mouthpart Types

Insect mouthparts, at least in the medically important species, can be generally divided into three broad categories:

1. Biting and chewing
2. Sponging
3. Piercing–sucking (Fig. 10.1)

Within these categories, there are numerous adaptations and/or specializations among the various insect orders. Biting and chewing mouthpart types, such as those

© Springer International Publishing AG, part of Springer Nature 2018
J. Goddard, *Infectious Diseases and Arthropods*, Infectious Disease,
https://doi.org/10.1007/978-3-319-75874-9_10

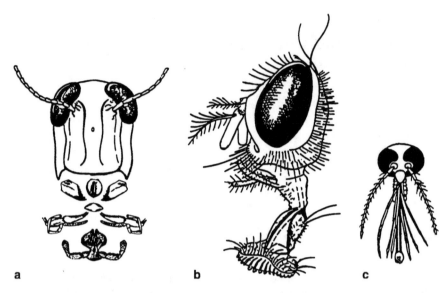

a                              b                              c

**Fig. 10.1** Various insect mouthpart types: (**a**) chewing, (**b**) sponging, and (**c**) piercing–sucking mouthparts. (Adapted from: US DHHS, CDC, Publication No. 83: 8297 and other sources)

in food pest insects, and sponging mouthpart types (Fig. 10.2d), found in the filth fly groups, are of little significance regarding human bites, but piercing–sucking mouthparts, and especially the bloodsucking types, are of considerable importance. Insect piercing–sucking mouthparts vary in the number and arrangement of needlelike blades (stylets) and the shape and position of the lower lip of insect mouthparts, the labium (Fig. 10.2). Often, what is termed the proboscis of an insect with piercing–sucking mouthparts is an ensheathment of the labrum, stylets, and labium. These mouthparts are arranged in such a way that they form two tubes. One tube is usually narrow, being a hollow pathway along the hypopharynx, and the other is wider, formed from the relative positions of the mandibles or maxillae. On biting, saliva enters the wound via the narrow tube, and blood returns through the wider tube by action of the cibarial or pharyngeal pump.

## 10.2.2  Sting Apparatus

In all stinging wasps, bees, and ants (insect order Hymenoptera), the stinger is a modified ovipositor, or egg-laying device, that usually no longer functions in egg laying. Accordingly, in the highly social Hymenoptera, only a queen or other repro- ductive caste member lays eggs; the workers gather food, conduct other tasks, and can sting intruders. They build paper nests under or above ground (Figs. 10.3 and 10.4) which they will aggressively defend. A typical ovipositor (nonstinging) con- sists of three pairs of elongate structures, called valves, which can insert the eggs

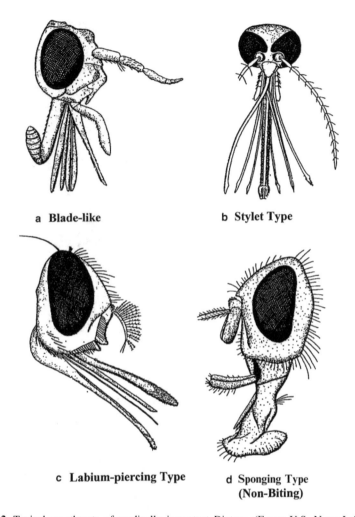

a  **Blade-like**                          b  **Stylet Type**

c  **Labium-piercing Type**        d  **Sponging Type**
                                          **(Non-Biting)**

**Fig. 10.2** Typical mouthparts of medically important Diptera. (From: U.S. Navy Laboratory Guide to Medical Entomology, 1943)

into plant tissues, soil, and so forth. One pair of the valves makes up a sheath and is not a piercing structure, whereas the other two pairs form a hollow shaft that can pierce substrate in order for the eggs to pass down through. Two accessory glands within the body of the female inject secretions through the ovipositor to coat the eggs with a glue-like substance.

For the stinging configuration, the ovipositor is modified to enable stinging (Fig. 10.5). The genital opening from which the eggs pass is anterior to the sting apparatus, which is flexed up out of the way during egg laying. Also, the accessory glands have been modified. One now functions as a venom gland, and the other, called the Dufour's gland, is important in the production of pheromones. The venom gland is connected to a venom reservoir or poison sac, which may contain up to 0.1 mL of venom in some of the larger hymenopterans.

**Fig. 10.3** Paper wasp
nests. (Photo copyright
2010 by Jerome Goddard,
Ph.D.)

**Fig. 10.4** Hornet's nest,
cut away to show inside.
(Photo copyright 2010 by
Jerome Goddard, Ph.D.)

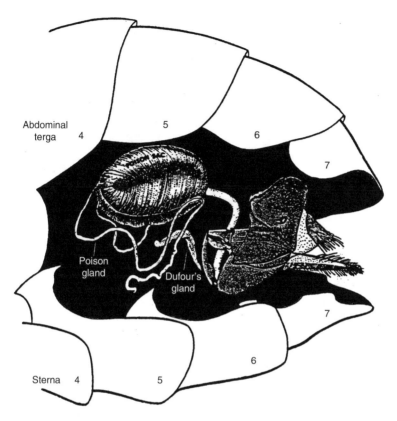

**Fig. 10.5** Cut away view of yellowjacket stinger. (USDA Agri. Handbk)

The stinger itself is well adapted for piercing vertebrate skin. In the case of yellowjackets (Fig. 10.6), there are two lancets and a median stylet that can be extended and thrust into a victim's skin. Penetration is not a matter of a single stroke but, instead, alternate forward strokes of the lancets, sliding along the shaft of the stylet. The tips of the lancets are slightly barbed (and actually recurved like a fishhook in the case of honeybees) so that they are essentially sawing their way through the victim's skin. Contraction of venom sac muscles injects venom through the channel formed by the lancets and shaft. The greatly barbed tip of the lancets in honeybees prevents the stinger from being withdrawn from vertebrate skin. Thus, the sting apparatus is torn out as the bee flies away. Other hymenopterans, on the other hand, can sting repeatedly.

## 10.2.3   Direct Damage to Tissue

Some lesions are the result of direct tissue damage from stings or bites. Arthropod mouthparts puncture skin by various mechanisms (siphoning tube, scissorlike blades, and so on) leading to skin damage. In this case, damage may be a small punctum, dual

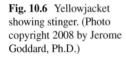

**Fig. 10.6** Yellowjacket showing stinger. (Photo copyright 2008 by Jerome Goddard, Ph.D.)

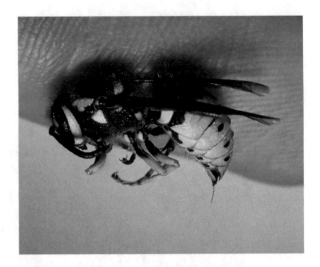

puncta (from fangs), or lacerations. By far, most lesions on the human skin are produced by host immune reactions to the offending arthropod salivary secretions or venom. Arthropod saliva is injected while feeding to lubricate the mouthparts on insertion, increase blood flow to the bite site, inhibit coagulation of host blood, anesthetize the bite site, suppress the host's immune and inflammatory responses, and/or aid in digestion. Stingers are needlelike structures that may puncture and damage human skin as well. Venom from certain spiders may directly affect the human skin, causing tissue death (necrosis). In the United States, violin spiders are primarily responsible for necrotic skin lesions, although sac spiders (*Cheiracanthium* spp.) and hobo spiders may also cause necrotic arachnidism [10, 11]. Brown recluse spider venom contains a lipase enzyme, sphingomyelinase D, which is significantly different from phospholipase A in bee and wasp venoms. This specific lipase is the primary necrotic agent involved in the formation of the typical lesions. It is possible that neutrophil chemotaxis is induced by sphingomyelinase D. The subsequent influx of neutrophils into the area is critical in the formation of the necrotic lesion (see Chap. 12).

## *10.2.4   Infectious Complications*

Secondary infection with common bacterial pathogens can occur in any lesion in which the integrity of the dermis is disrupted, whether by necrosis or excoriation [12]. Infection may result in cellulitis, impetigo, ecthyma, folliculitis, furunculosis, and other manifestations. Three findings may be helpful in making the diagnosis of secondary bacterial infection [12]:

1. Increasing erythema, edema, or tenderness beyond the anticipated pattern of response of an individual lesion suggests infection.
2. Regional lymphadenopathy can be a useful sign of infection, but it may also be present in response to the primary lesion without infection.
3. Lymphangitis is the most reliable sign and suggests streptococcal involvement.

## 10.3  Clues to Recognizing Insect Bites or Stings

### 10.3.1  *Diagnosis*

If a patient recalls no insect or arachnid exposure, arthropod bites or stings may pose difficulty in diagnosis. No physician or entomologist can accurately determine what insect caused a particular bite or sting lesion; however, there might be helpful clues. Alexander [1] described a typical hymenopteran sting (excluding ants) as a central white spot marking the actual sting site surrounded by an erythematous halo. Generally, the entire lesion is a few square centimeters in area. Of course, allergic reactions may result in much larger lesions. He also described an initial rapid dermal edema with neutrophil and lymphocyte infiltration. Plasma cells, eosinophils, and histiocytes appear later.

Arthropod bites should be considered in the differential diagnosis of any patient complaining of itching. Bites are characterized by urticarial wheals, papules, vesicles, and less commonly, blisters. After a few days or even weeks, secondary infection, discoloration, scarring, papules, or nodules may persist at the bite site. Complicating the picture further is the development of late cutaneous allergic responses in some atopic individuals. Diagnosis may be especially difficult in the case of biopsies of papules or nodules. Biopsies may reveal a dense infiltrate of a mixture of inflammatory cells, such as lymphocytes, plasma cells, histiocytes, giant cells, neutrophils, and eosinophils. Lesions containing a majority of lymphocytes could be mistaken for a lymphomatous infiltrate. If the infiltrate is predominantly perivascular and extending throughout the depths of the dermis, the lesion might be confused with a lupus erythematosus. Eosinophils are commonly seen in papules or nodules from arthropod bites. There may be a dense infiltration of neutrophils, resembling an abscess. Occasionally arthropod mouthparts may still be present within the lesion, and there may be a granulomatous inflammation in and around these mouthparts. Scabies mites occur in the stratum corneum and can usually be seen on microscopic examination. New lesions from scabies, such as papules or vesicles are covered by normal keratin, whereas older lesions have a heaped-up parakeratotic surface. There may also be a perivascular infiltrate of lymphocytes, histiocytes, and eosinophils. Histopathologic studies of late cutaneous allergic responses have revealed mixed cellular infiltrates, including lymphocytes, polymorphonuclear leukocytes, and some partially degranulated basophils. A prominent feature of late cutaneous allergic reactions has been fibrin deposition interspersed between collagen bundles in the dermis and subcutaneous tissues.

Diagnosis of insect bites or stings depends on:

1. Maintaining a proper index of suspicion in this direction (especially during the summer months)
2. A familiarity of the insect fauna in one's area
3. Obtaining a good history

It is very important to find out what the patient has been doing lately, e.g., hiking, fishing, gardening, cleaning out a shed, and so forth. However, even history can be misleading in that patients may present with a lesion that they think is a bite or sting, when in reality the correct diagnosis is something like urticaria, folliculitis, or delusions of parasitosis (see Chap. 14). Physicians need to be careful not to diagnose "insect bites" based on lesions alone and should call on entomologists to examine samples.

## 10.4   Summary and Conclusions

A human's first line of defense against invasion or external stimuli is the skin. It may react in a variety of ways against all kinds of stimuli—physical or chemical—including arthropods and their emanations. Lesions may result from arthropod exposure, although not all lesions have the same pathological origin—some are owing to mechanical trauma, some owing to infectious disease processes, and some result from sensitization processes. Physicians and other healthcare providers are frequently confronted with patients having skin lesions attributed to a mysterious arthropod bite or sting. Diagnosis is difficult but may be aided by asking the patient numerous questions about the event and any recent activity that might have led to arthropod exposure. The following questions might provide useful information: "Did you see the offending arthropod?" "Was it wormlike?" "Did it fly?" "Where were you when these lesions occurred?" Most treatments (except in cases of infectious diseases) involve counteracting immune responses to venoms, salivary secretions, or body parts using various combinations of antihistamines and corticosteroids. Infectious diseases often require antibiotic/supportive care.

## References

1. Alexander JO. Arthropods and human skin. Berlin: Springer; 1984. p. 422.
2. Allington HV, Allington RR. Insect bites. J Am Med Assoc. 1954;155:240–7.
3. Frazier CA. Diagnosis of bites and stings. Cutis. 1968;4:845–9.
4. Goddard J. Physician's guide to arthropods of medical importance. 6th ed. Boca Raton: Taylor and Francis (CRC); 2013. p. 412.
5. O'Neil ME, Mack KA, Gilchrist J. Epidemiology of non-canine bite and sting injuries treated in U.S. emergency departments, 2001-2004. Pub Health Rep. 2007;122:764–75.
6. Goddard J. Direct injury from arthropods. Lab Med. 1994;25:365–71.
7. Lehman CF, Pipkin JL, Ressmann AC. Blister beetle dermatitis. Arch Dermatol. 1955;71:36–41.
8. Shpall S, Freiden I. Mahogany discoloration of the skin due to the defensive secretion of a millipede. Pediatr Dermatol. 1991;8:25–6.
9. Rolla G, Franco N, Giuseppe G, Marsico P, Riva G, Zanotta S. Cotton wool in pine trees. Lancet. 2003;361:44.
10. CDC. Necrotic arachnidism -- Pacific northwest, 1988-1996. CDC. MMWR. 1996;45:433–6.
11. Diaz JH. The global epidemiology, syndromic classification, management, and prevention of spider bites. Am J Trop Med Hyg. 2004;71(2):239–50.
12. Kemp ED. Bites and stings of the arthropod kind. Postgrad Med. 1998;103:88–94.

# Chapter 11
# Fire Ant Attacks on Humans

## 11.1 The Problem

Ants (insect family Formicidae) are perhaps the most successful organisms on earth. Their highly varied lifestyles and social structure make them an interesting and much-studied group [1]. Almost all ant species can bite humans, and some species sting. Some notorious stinging ants include the famous bulldog ants in Australia, bullet ants in South America, harvester ants in the western United States, and imported fire ants, also in the United States. During the past 25 years, a series of articles was published describing the biologic and entomologic characteristics of imported fire ants (IFAs) and the medical consequences of their stings on humans [2–6]. IFAs include *Solenopsis invicta*, *Solenopsis richteri*, and their hybrid *Solenopsis invicta* × *richteri* [7, 8]. Fire ants currently infest at least 350 million acres over much of the southern United States from coast to coast. Because of the ubiquity and aggressiveness of the ants, human encounters with IFAs in areas in which the ants are endemic are virtually inevitable. The ants favor disturbed habitats, and the progressive urbanization of the United States, especially in the Sun Belt, has accelerated their spread.

Polygyne (multiple queen) organization, in which numerous egg-laying queens reside in a single colony, permits as many as 600 or so fire ant mounds per acre in some areas. In rural areas, fire ants may attack both humans and animals [5]. They also may damage farm equipment, electrical systems, irrigation systems, and crops. In urban settings, fire ants build mounds in sunny, open areas, such as lawns, playgrounds, ball fields, parks, and golf courses, as well as along road shoulders and median strips. This brings them into close contact with humans. Making matters worse, sometimes they will build nests along the foundation walls of houses and other dwellings.

IFAs seek sites necessary for colony survival during periods of environmental stress, such as during food shortages, hot and dry summers, or heavy rainfall. Inhabited dwellings can be ideal environments for fire ants because of the

© Springer International Publishing AG, part of Springer Nature 2018
J. Goddard, *Infectious Diseases and Arthropods*, Infectious Disease,
https://doi.org/10.1007/978-3-319-75874-9_11

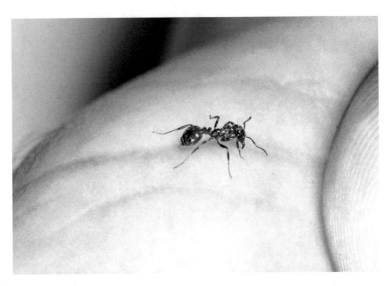

**Fig. 11.1** Fire ant stinging. (Photo courtesy Dr. James Jarratt, Mississippi State University Extension Service)

availability of food, moisture, and protection from extremes in weather. Thus, humans come into contact with ants not only outdoors but also indoors.

When a mound is disturbed, thousands of ants swarm to the surface and sting just about anything in sight. Few, if any, personal protective measures have any effect against fire ant attacks. This author has conducted numerous experiments testing repellents and other chemical substances against fire ants to see whether any of these substances stopped (or even slowed) the stinging response; nothing seemed to work [9]. In experiments using a twisted paper towel saturated with various substances, sulfur was slightly more effective in mitigating the sting response than other substances. However, when a child's sock was saturated with sulfur and placed in a fire ant mound, there was rapid and aggressive stinging of the sock.

Typically, 30–60% of persons living in infested urban areas are stung by IFAs each year [3, 10, 11]. However, one survey reported stings in 89% of subjects or immediate family members per year [12]. Furthermore, 55 (51%) of 107 previously unexposed persons were stung within 3 weeks of arrival in an area in which fire ants were endemic, and specific IgE antibody to fire ant venom developed in 8 (14.5%) of these 55 [13].

Stings occur most frequently during summer, most commonly in children, and typically on the lower extremities. When stinging, the ant uses its powerful mandibles to hold onto the skin, often arching its body, and injects venom through the stinger located at the tip of its abdomen (Fig. 11.1). The ant will sting repeatedly if not quickly removed. Stings are characterized by an immediate intense burning (the "fire" inspiring the name of the ant) and itching at the sting site. However, stings that occur during the off-season (winter months) may not cause

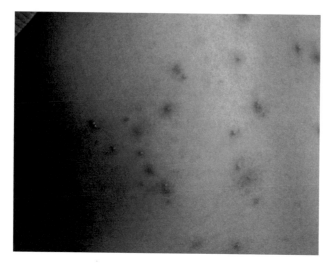

**Fig. 11.2** Fire ant stings on back. (Photo copyright 2005 by Jerome Goddard, Ph.D.)

as much pain and may go unnoticed until the local reaction develops. This may reflect seasonal differences in IFA venom protein concentration. Generally, within 8–24 h, a pustule develops at each sting site which may persist 3 d to 1 wk [14] (Fig. 11.2).

## 11.2 Effects of the Venom

Fire ant venom is different from venom of most other stinging insects because it contains only 1% protein [15]. The venom possesses hemolytic, neurotoxic, and cytotoxic activity and has the ability to inhibit sodium and potassium adenosine triphosphatases, reduce mitochondrial respiration, uncouple phosphorylation, and adversely affect neutrophil and platelet function [16, 17]. Also, it has been found to inhibit nitric oxide synthetase [17, 18]. Nitric oxide inhibitors may promote bronchospasm during anaphylaxis and adversely affect cardiac function [19]. The presence of D-dimers noted in some patients who have been stung reflects activation of the contact system by venom components [4, 17, 20].

All the above properties of fire ant venom may contribute to activation of the coagulation system and severity of anaphylaxis seen in some patients [4, 21]. When symptoms compatible with acute allergic reactions develop in stung patients, serum tryptase levels may be useful in distinguishing anaphylaxis from other reactions [6, 21]. In a previous study, serum tryptase levels obtained within 24 h of death were elevated in nine of nine persons who died of anaphylaxis after hymenopteran stings [22]. During an anaphylactic episode, levels generally reach a peak 15–30 min after the sting and then decrease, with a half-life of 1.5–2.5 h.

## 11.3   Infectious Complications

Secondary complications from fire ant stings frequently involve bacterial infections. Chronically ill or intoxicated patients who receive many stings are the persons in whom complications most commonly occur [23]. The originally sterile fluid in pustules becomes contaminated after a person scratches the lesions. Although infections are usually not very severe, generalized sepsis and renal insufficiency have been observed [24]. Three case reports of secondary infection have been reported in detail; one involved cellulitis from β-hemolytic streptococci [23, 25]. For the prevention of such infections, the authors of that study recommended cleansing sting sites with soap and water, avoiding excoriation, and immediately treating secondary infection [24, 25]. In addition to bacterial infections, there is at least one report of a fungal infection (sporotrichosis) resulting from fire ant stings [25].

## 11.4   Protecting Patients in Healthcare Facilities
##          from Ant Attacks

The extremes in weather that cause movement of fire ants into inhabited dwellings are especially problematic for healthcare facilities, such as nursing homes. During the spring, when soils become saturated, IFA colonies may move inside to look for drier conditions. Similar movement of ants may occur during periods of drought, when they will travel toward moisture if it is found inside. Another important factor facilitating movement of IFAs to the inside is proximity of ant mounds to the foundation of a building.

Most persons are able to detect fire ants' stinging and thus can move, jump, or run to avoid further injury, but special care is required to ensure that patients in long-term care facilities are not stung by fire ants [26, 27]. Patients in these facilities may not be aware of their surroundings, may be immobilized by disease, or may be otherwise incapacitated and unable to respond if ants come in contact with them. Once foraging fire ants come in contact with a patient, a variety of external stimuli, including movement of the patient, might trigger a stinging event that leads to multiple stings in a very short period.

Some commonsense suggestions for prevention of indoor fire ant infestations include:

1. Watching for IFA infestations indoors during weather extremes
2. Keeping patients' beds and linens away from walls and floors
3. Limiting food in beds
4. Placing food in the room in a well-sealed (airtight) container

Fire ant management also includes a systematic plan for keeping the pests out of healthcare facilities and, if they enter, ways to mitigate their effects. Close coordination with a licensed pest control firm is critical. Once fire ants are found on a patient, clinical evaluation is needed as well as possible transport (depending on findings) to the nearest emergency department.

# References

1. Holldobler B, Wilson EO. The ants. Cambridge, MA: Harvard University Press; 1990. p. 732.
2. de Shazo RD, Williams DF. Multiple fire ant stings indoors. South Med J. 1995;88:712–5.
3. de Shazo RD, Butcher BT, Banks WA. Reactions to the stings of the imported fire ant. N Engl J Med. 1990;323:462–6.
4. de Shazo RD, Kemp SF, deShazo MD, Goddard J. Fire ant attacks on patients in nursing homes: an increasing problem. Am J Med. 2004;116(12):843–6.
5. Goddard J, de Shazo RD. Fire ant attacks on humans and animals. Encyclopedia of Pest Management (online), 120024662; 2004. https://doi.org/10.1081/E-EPM.
6. Kemp SF, deShazo RD, Moffitt JE, Williams DF, Buhner WA 2nd. Expanding habitat of the imported fire ant (Solenopsis invicta): a public health concern. J Allergy Clin Immunol. 2000;105(4):683–91.
7. Buren WF. Revisionary studies on the taxonomy of the imported fire ants. J Georgia Entomol Soc. 1972;7:1–25.
8. Trager JC. A revision of the fire ants, Solenopsis geminata group. J New York Entomol Soc. 1991;99:141–98.
9. Goddard J. Personal protection measures against fire ant attacks. Ann Allergy Asthma Immunol. 2005;95:344–9.
10. de Shazo RD, Griffing C, Kwan T, Banks WA, Dvorak HF. Dermal hypersensitivity reactions to imported fire ants. J Allergy Clin Immunol. 1984;74:841–7.
11. Vinson SB. Invasion of the red imported fire ant. Am Entomol. 1997;43:23–39.
12. Tracy JM, Demain JG, Quinn JM, Hoffman DR, Goetz DW, Freeman T. The natural history of exposure to the imported fire ant. J Allergy Clin Immunol. 1995;95:824–8.
13. Hoffman DR, Dove DE, Jacobson RS. Allergens in hymenoptera venom. XX. Isolation of four allergens from imported fire ant venom. J Allergy Clin Immunol. 1988;82:818–21.
14. Goddard J, Jarratt J, de Castro FR. Evolution of the fire ant lesion. J Am Med Assoc. 2000;284:2162–3.
15. Jones TH, Blum MS, Fales HM. Ant venom alkaloids from Solenopsis and Monomovian species. Tetrahedron. 1982;38:1949–58.
16. Javors MA, Zhou W, Maas JW Jr, Han S, Keenan RW. Effects of fire ant venom alkaloids on platelet and neutrophil function. Life Sci. 1993;53(14):1105–12.
17. Yi GB, McClendon D, Desaiah D, Goddard J, Lister A, Moffitt J, et al. Fire ant venom alkaloid, isosolenopsin A, a potent and selective inhibitor of neuronal nitric oxide synthase. Int J Toxicol. 2003;22(2):81–6.
18. Mitsuhata H, Shimizu R, Yokoyama MM. Role of nitirc acid in anaphylactic shock. J Clin Immunol. 1995;15:277–83.
19. de Shazo RD, Banks WA. Medical consequences of multiple fire ant stings occurring indoors. J Allergy Clin Immunol. 1994;93:847–50.
20. Schwartz LB, Metcalfe DD, Miller JS, Earl H, Sullivan T. Tryptase levels as an indicator of mast-cell activation in systemic anaphylaxis and mastocytosis. N Engl J Med. 1987;316(26):1622–6.
21. Yunginger JW, Nelson DR, Squillace DL. Laboratory investigation of deaths due to anaphylaxis. J Forensic Sci. 1991;36:857–65.
22. Cohen PR. Imported fire ant stings: clinical manifestations and treatment. Pediatr Dermatol. 1992;9(1):44–8.
23. Stablein JJ, Lockey RF. Adverse reactions to ant stings. Clin Rev Allergy. 1987;5:161–75.
24. Parrino J, Kandawalla NM, Lockey RF. Treatment of local skin response to imported fire ant stings. South Med J. 1981;74:1361–4.
25. Miller SD, Keeling JH. Ant sting sporotrichosis. Cutis. 2002;69:439–42.
26. Goddard J. New record for Ixodes texanus banks in Mississippi, with a new host record. Entomol News. 1983;94:139–40.
27. Goddard J, Jarratt J, deShazo RD. Recommendations for prevention and management of fire ant infestation of health care facilities. South Med J. 2002;95(6):627–33.

# Chapter 12
# Necrotic Arachnidism: Brown Recluse Bites

## 12.1 Introduction and Biology

Spiders in the family Loxoscelidae, and specifically the genus *Loxosceles* (comprising more than 50 species in Eurasia, Africa, and the Americas), are medically important because of their cytotoxic and hemolytic venom which may cause significant tissue necrosis [1]. In the United States, the most notorious member of this genus is *Loxosceles reclusus*, the brown recluse (BR) [2], although several other *Loxosceles* species live in the southwestern US deserts (Fig. 12.1) [2, 3]. Many cases of necrotic skin wounds—necrotic arachnidism—have been attributed to bites by these spiders [3–5], as have fatalities due to systemic reactions, such as hemolytic anemia [6–8].

The biologic characteristics and distribution of the BR spider have been described elsewhere [9, 10]. Adult BR spiders are about the size of a quarter (legs included) and may be any one of several shades of brown. There are no visible markings other than a well-defined dark area on the cephalothorax that resembles a violin—hence, the common name, fiddleback spider (Fig. 12.2). These spiders are reclusive, preferring dark areas for their habitat, such as attics, closets, basements, sheds, barns, and other outbuildings. They may also be found behind pictures of furniture or in stacks of papers, debris, and firewood in and around the home.

BR spiders spin a coarse, irregular web and, unlike orb-weaving spiders, do not produce the large, prominent webs commonly seen around homes. The endemic range of the BR spider is southeastern Nebraska through Texas, east to Georgia, and southernmost Ohio. These spiders may occasionally be found in homes or household goods in areas in which they are not endemic when people move from one part of the country to another.

© Springer International Publishing AG, part of Springer Nature 2018
J. Goddard, *Infectious Diseases and Arthropods*, Infectious Disease,
https://doi.org/10.1007/978-3-319-75874-9_12

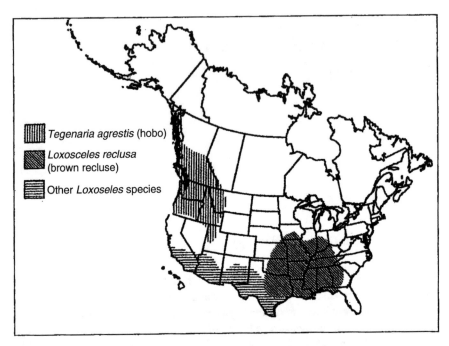

**Fig. 12.1** Approximate geographic distributions of spiders that cause necrotic arachnidism. (CDC figure)

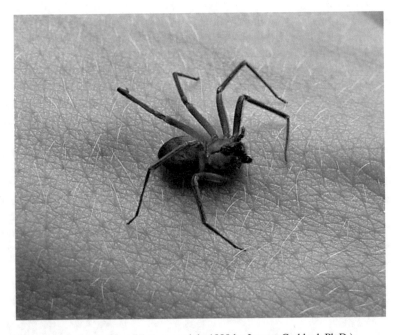

**Fig. 12.2** Brown recluse spider. (Photo copyright 1999 by Jerome Goddard, Ph.D.)

## 12.2   Facts and Fables About Brown Recluse Bites

Although the bites of BR spiders are dangerous, the evidence for these spiders' aggressive biting behavior and the widespread negative health impact on humans is not as strong as once thought. Almost 5000 BR spider bites are reported to poison control centers each year; if epidemiologic data and confirmed cases are any indication, most of the reported bites are caused by something else [11, 12]. The problem is that hundreds of BR spider bites are being diagnosed in areas in which the spiders occur sparsely, if at all [13]. For example, in 41 mo of data collection, researchers were informed of 216 BR spider bite diagnoses from California, Oregon, Washington, and Colorado; however, the same researchers could only confirm identification of 35 BR or Mediterranean recluse spiders from those same four states [14]. In Florida, medical personnel diagnosed 124 BR spider bites from 31 counties during a 6-yr period, yet arachnologists only have records of about 70 BR spiders being found in ten Florida counties over the past 100 yr. [15]!

I personally am aware of two correctional facilities in Mississippi where "dozens" of medically documented BR spider bites supposedly occurred, yet pest-control professionals and Extension Service entomologists failed to find even one BR spider. From the entomologic perspective, it borders on the ridiculous to claim that dozens of persons have been bitten by BR spiders when none can be found. These spiders are not that reclusive.

BR spiders are not aggressive. A BR spider typically bites defensively when it is accidentally trapped against human skin while a person is dressing or sleeping. Often, no biting incidents occur even when hundreds of the spiders are present in a dwelling. Vetter and Barger [16] reported finding more than 2000 BR spiders in a home in Kansas inhabited by a family of four, with no bites recalled or reported by the family.

Necrotic reactions to BR spider bites are also probably exaggerated. Things must be kept in perspective. Certainly, there is evidence that the venom can cause unsightly spots of necrosis on the human skin. As a graduate student, I participated in experiments in which laboratory animals were injected intradermally with BR spider venom. Horrible lesions often developed within 10 d of those injections, so no one is going to convince me that these lesions cannot also happen on the human skin.

Reactions to BR spider bites can vary from no reaction to a mild red wound to a wound with terrifying necrotic flesh. However, Masters and King [17] say that cutaneous necrosis usually does not develop after untreated BR spider bites. Cacy and Mold [18] reported that 149 of 405 BR spider bites produced necrosis, which means that 60% of bites did not. Therefore, the majority of BR spider bites heal on their own without serious scarring.

Reports of pain from BR spider bites have been well-supported. Older studies often noted pain in BR spider bite cases [18], but a careful study of 46 well-documented cases showed that pain (often severe) results from BR bites [19].

Reports of deaths from BR spider bites are not strongly supported. To my knowledge, no deaths from BR spider bites have been proved—cases in which the offending spider was collected and identified by an expert. Again, things must be kept in perspective. Most claims of deaths from BR spider bites (caused by subsequent hemolytic anemia) are reasonable, intuitive, and probably true, but absolute proof is lacking.

## 12.3   Differential Diagnosis

The diagnosis of BR spider bites is based on clinical presentation and history (unless, of course, the patient brings in the offending specimen). Although a diagnostic ELISA test has been developed [20, 21], it is considered experimental and not widely available at this time.

The BR spider bite lesion varies in appearance, but generally it is a dry blue-gray or blue-white irregular sinking patch with ragged edges, surrounded by erythema (the "red, white, and blue sign"). The lesion often is asymmetric and gravity-dependent, because of the downward flow of venom through tissues (Fig. 12.3). A perfectly symmetric necrotic lesion is often not caused by a BR spider bite. Necrosis and sloughing of tissues may follow over days or weeks, leading to an unsightly, sunken scar (Fig. 12.4).

Many other conditions may lead to spots of necrosis on the human skin resembling BR spider bites and should be considered by physicians before they make a diagnosis of BR spider bite (Table 12.1). The most common causes of necrotic wounds misdiagnosed as BR spider bites are infections with *Staphylococcus* and *Streptococcus* species. Skin and soft tissue infections caused by methicillin-resistant *Staphylococcus aureus* are steadily increasing [22].

## 12.4   Treatment of Bites

Unfortunately, fables also plague the subject of treating patients who have BR spider bites. The fact that a large proportion of bites are unremarkable and do not become necrotic contributes to the confusion. For example, a layperson might claim that rubbing tobacco juice on a bite lesion prevents necrosis (because no necrosis subsequently occurs). Before long, the idea of using tobacco juice for BR spider bites would become widespread.

There are several treatments for BR spider bites reported in the lay and scientific literature, such as excision, systemic or locally injected corticosteroids, antibiotics, antihistamines, colchicines, electric shock, hyperbaric oxygen, and nitroglycerin patches of ointment [18, 23–26]. However, controlled studies of these strategies are mostly lacking. Evidence indicates that surgical excision is not beneficial and, in most cases, delays healing [27]. Systemic corticosteroids have also been associated with slower healing of BR spider bites [28]. Nitroglycerin apparently does not help [29]. One randomized controlled study failed to show any benefit of hyperbaric oxygen in the management of BR spider bites [30]. The leukocyte inhibitor dapsone has often been recommended for management of BR spider bites, but there is evidence that dapsone may not be effective in treating BR spider bites. A randomized, blinded, controlled study of venom effects in rabbits failed to show any benefit from

**Fig. 12.3** Brown recluse
spider bite showing
asymmetric shape. (CDC
photo)

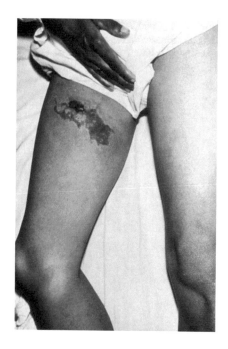

**Fig. 12.4** Brown recluse
spider bite 4 months
post-bite. (Photo from
Mississippi State
University Extension
Service, Publ. No. 2154)

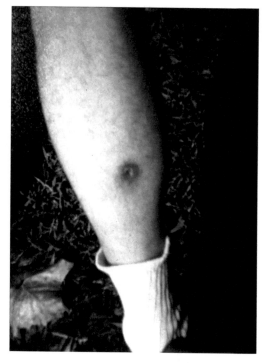

**Table 12.1** Conditions that
may be mistaken for brown
recluse spider bite[a]

| |
| --- |
| Anthrax |
| Cellulitis |
| Chagas' disease |
| Contact or chemical dermatitis |
| Cutaneous/focal vasculitis |
| Decubitus ulcer |
| Diabetic or venous stasis ulcer |
| Drug eruption |
| Ecthyma gangrenosum |
| Herpes simplex |
| Herpes zoster |
| Impetigo |
| Lyme disease |
| Lymphomatoid papulosis |
| Necrotizing fasciitis |
| Other arthropod bites |
| Polyarteritis nodosa |
| Pyoderma gangrenosum |
| Soft tissue trauma |
| Sporotrichosis |
| Tularemia |

[a]Not an exhaustive list of conditions that may cause necrotic wounds

the use of this drug [31]. Another study showed that use of dapsone was associated with slower healing of BR spider bite wounds [28]. Therefore, the best treatment strategy for BR spider bites is supportive treatment only [32]. Most wounds heal without intervention, although a scar may remain [32].

# References

1. Lane RP, Crosskey RW. Medical insects and arachnids. New York: Chapman and Hall; 1996. p. 723.
2. Goddard J. Physician's guide to arthropods of medical importance. 6th ed. Boca Raton: Taylor and Francis (CRC); 2013. p. 412.
3. Russell FE, Waldron WG, Madon MB. Bites by the brown spiders *Loxosceles unicolor* and *Loxosceles arizonica* in California and Arizona. Toxicon. 1969;7:109–17.
4. Lessenden CM, Zimmer LK. Brown spider bites. J Kans Med Soc. 1960;61:379–85.
5. Russell FE, Wainsschel J, Gertsch WJ. Bites of spiders and other arthropods. In: Conn HF, editor. Current therapy. Philadelphia: W.B. Saunders; 1973. p. 868.
6. Hostetler MA, Dribben W, Wilson DB, Grossman WJ. Sudden unexplained hemolysis occurring in an infant due to presumed *Loxosceles* envenomation. J Emerg Med. 2003;25:277–82.
7. Leung LK, Davis R. Life-threatening hemolysis following a brown recluse spider bite. J Tenn Med Assoc. 1995;32:396–7.

8. Murray LM, Seger DL. Hemolytic anemia following a presumptive brown recluse spider bite. Clin Toxicol. 1994;32:451–6.
9. CDC. Necrotic arachnidism -- Pacific northwest, 1988-1996. CDC. MMWR. 1996;45:433–6.
10. Hite JM, Gladney WJ, Lancaster JLJ, Whitcomb WH. Biology of the brown recluse spider. Fayetteville: University of Arkansas Agricultural Experiment Station. Bull. No. 711; 1966.
11. O'Neil ME, Mack KA, Gilchrist J. Epidemiology of non-canine bite and sting injuries treated in U.S. emergency departments, 2001-2004. Pub Health Rep. 2007;122:764–75.
12. Sandlin N. Convenient culprit: myths surround the brown recluse spider. Amednews.com (e-news for America's physicians). Chicago: American Medical Association. Accessed 5 Aug 2002.
13. Vetter RS, Bush SP. The diagnosis of brown recluse spider bite is overused for dermonecrotic wounds of uncertain etiology. Ann Emerg Med. 2002;39:544–6.
14. Vetter RS, Cushing PE, Crawford RL, Royce LA. Diagnoses of brown recluse spider bites greatly outnumber actual verifications of the spider in four western American states. Toxicon. 2003;42:413–8.
15. Vetter RS, Edwards GB, James LF. Reports of envenomation by brown recluse spiders outnumber verifications of Loxosceles spiders in Florida. J Med Entomol. 2004;41:593–7.
16. Vetter RS, Barger DK. An infestation of 2,055 brown recluse spiders and no envenomation in a Kansas home: implications for bite diagnosis in nonendemic areas. J Med Entomol. 2002;39:948–51.
17. Masters EJ, King LE Jr. Differentiating loxoscelism from Lyme disease. Emerg Med. 1994;26:47–9.
18. Cacy J, Mold JW. The clinical characteristics of brown recluse spider bites treated by family physicians. J Fam Pract. 1999;48:536–42.
19. Payne KS, Schilli K, Meier K, Rader RK, Dyer JA, Mold JW, et al. Extreme pain from brown recluse spider bites: model for cytokine-driven pain. JAMA Dermatol. 2014;150(11):1205–8.
20. Gomez HF, Krywko DM, Stoecker WV. A new assay for the detection of Loxosceles species spider venom. Ann Emerg Med. 2002;39:469–74.
21. Stoecker WV, Green JA, Gomez HF. Diagnosis of loxoscelism in a child confirmed with an enzyme-linked immunosorbent assay and noninvasive tissue sampling. J Am Acad Dermatol. 2006;55:888–90.
22. Johnson JK, Khole T, Shurland S, Kereisel K, Stine OC, Roghmann MC. Skin and soft tissue infections caused by methicillin-resistant Staphylococcus aureus USA300 clone. Emerg Infect Dis. 2007;13:1195–9.
23. King LE Jr, Rees RS. Dapsone treatment of a brown recluse bite. J Am Med Assoc. 1983;250:648.
24. Masters EJ. Loxoscelism. N Engl J Med. 1998;339:379.
25. Rees R, Campbell D, Rieger E, King LE. The diagnosis and treatment of brown recluse spider bites. Ann Emerg Med. 1987;16:945–9.
26. Rees RS, Altenbern DP, Lynch JB, King LE Jr. Brown recluse spider bites: a comparison of early surgical excision versus dapsone and delayed surgical excision. Ann Surg. 1985;202:659–63.
27. Merigian KS, Blaho K. Envenomation from the brown recluse spider: review of mechanism and treatment options. Am J Ther. 1996;3:724–34.
28. Mold JW, Thompson DM. Management of brown recluse spider bites in primary care. J Am Board Fam Pract. 2004;17:347–52.
29. Lowry BP, Bradfield JF, Carroll RG. A controlled trial of topical nitroglycerin in a New Zealand white rabbit model of brown recluse spider envenomation. Ann Emerg Med. 2001;37:161–5.
30. Escalante-Galindo P, Montoya-Cabrera MA, Terroba-Larios VM. Local dermonecrotic loxoscelism in children bitten by the spider Loxosceles reclusa [in Spanish]. Gac Med Mex. 1999;135:423–6.
31. Phillips S, Kohn M, Baker D, Vander Leest R, Gomez H, McKinney P, et al. Therapy of brown spider envenomation: a controlled trial of hyperbaric oxygen, dapsone, and cyproheptadine. Ann Emerg Med. 1995;25:363–8.
32. Bope ET, Kellerman R. Conn's current therapy. Philadelphia: Elsevier Saunders; 2017. p. 1375.

# Chapter 13
# Myiasis

## 13.1 Introduction and Medical Significance

The condition of fly maggots infesting the tissues of people or animals is referred to as myiasis. Specific cases of myiasis are clinically defined by the affected areas(s) involved. For example, there may be traumatic (wound), gastric, rectal, auricular, and urogenital myiasis, among others. Although not an infectious disease in the strictest sense, myiasis cases are often seen by family physicians or infectious disease specialists. Myiasis can be accidental, when fly larvae occasionally find their way into the human body, or facultative, when fly larvae enter living tissue opportunistically after feeding on decaying tissue in neglected, malodorous wounds. Myiasis can also be obligate in which the fly larvae must spend part of their developmental stages in living tissue. Obligate myiasis is true parasitism and is the most serious form of the condition.

Fly larvae are not capable of reproduction, and therefore, myiasis should not be considered contagious from patient to patient. Transmission of myiasis occurs only via an adult female fly.

### 13.1.1 Accidental Myiasis

Accidental enteric myiasis (sometimes referred to as pseudomyiasis) is mostly a benign event, but fly larvae could possibly survive temporarily, causing stomach pains, nausea, or vomiting. However, care should be exercised in diagnosing enteric myiasis, since many cases, some of which get into the scientific literature, are actually contamination of the toilet bowl or stool itself after the fact. Seeing maggots in the stool or toilet bowl is so alarming that patients may overlook other possibilities. This author once investigated a case wherein soldier fly (*Hermetia illucens*) larvae were frequently being found in a woman's toilet bowl (Fig. 13.1). She, of course,

© Springer International Publishing AG, part of Springer Nature 2018
J. Goddard, *Infectious Diseases and Arthropods*, Infectious Disease,
https://doi.org/10.1007/978-3-319-75874-9_13

**Fig. 13.1** Soldier fly larvae which are often found in toilet bowls. (Photo copyright 2007 by Jerome Goddard, Ph.D.)

**Fig. 13.2** Soldier fly larva vomited up by a child. (Photo copyright 2014 by Jerome Goddard, Ph.D.)

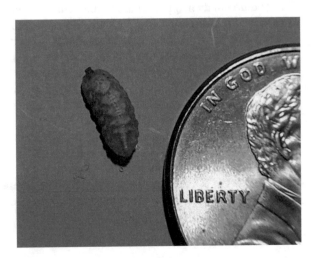

feared that the larvae were infesting her body. As it turned out, on disengaging and lifting the toilet up from the floor, numerous fly larvae were found living in the "scum" lining the pipe and even in the toilet wax seal.

Certainly some cases are genuine [1]. I have seen well-documented cases where the larvae were vomited up (Fig. 13.2) [2]. Numerous fly species in the families Muscidae, Calliphoridae, and Sarcophagidae may produce accidental enteric myiasis. Some notorious offenders are the cheese skipper, *Piophila casei* (Fig. 13.3); the black soldier fly, *H. illucens*; and the rat-tailed maggot, *Eristalis tenax* (Fig. 13.4). Other instances of accidental myiasis occur when fly larvae enter the urinary passages or other body openings. Flies in the genera *Musca, Muscina, Fannia, Megaselia* (Fig. 13.5), and *Sarcophaga* have often been implicated in such cases.

**Fig. 13.3** Cheese skipper
*P. casei.* (From: USDA,
publ. Ref. [5])

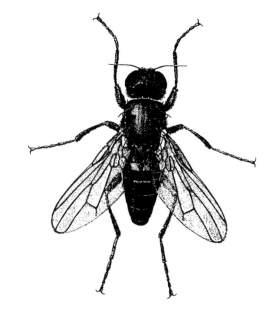

**Fig. 13.4** Rat-tailed
maggot *E. tenax* (**a**) adult
and (**b**) larva. (From
USDA publ., Ref. [5])

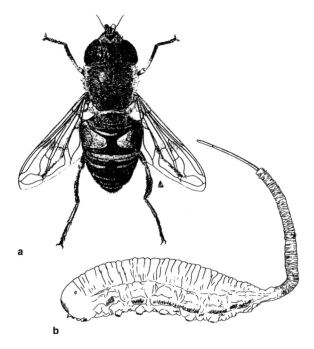

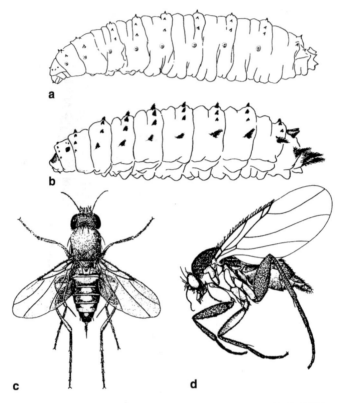

**Fig. 13.5** Various hump-backed flies and their larvae. (From USDA publ., Ref. [5])

## 13.1.2   Facultative Myiasis

Facultative myiasis may result in pain and tissue damage if fly larvae leave necrotic tissues and invade healthy tissues. Numerous species of Muscidae, Calliphoridae, and Sarcophagidae have been reported in cases of facultative myiasis (Figs. 13.6 and 13.7). In the United States, the calliphorid *Lucilia sericata* has been reported causing facultative myiasis on several occasions [3–5]. Another calliphorid, *Chrysomya rufifacies*, has been recently introduced into the United States from the Australasian region and is also known to be regularly involved in facultative myiasis [6]. Other muscoid fly species that may be involved in this type of myiasis include *Calliphora vicina*, *Phormia regina*, *Cochliomyia macellaria* (Fig. 13.8), and *Sarcophaga haemorrhoidalis*.

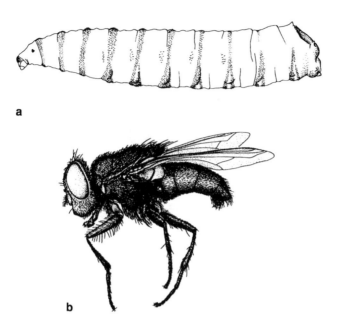

**Fig. 13.6**   Blowfly *C. macellaria*: (**a**) larva and (**b**) adult. (From USDA publ., Ref. [5])

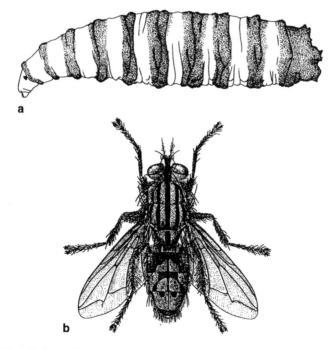

**Fig. 13.7**   Flesh fly *Sarcophaga* spp. (**a**) larva and (**b**) adult. (From USDA publ., Ref. [5])

**Fig. 13.8** The secondary screwworm, *Cochliomyia macellaria*. (Photo copyright 2014 by Jerome Goddard, Ph.D.)

### 13.1.3    Obligate Myiasis

Some fly species must develop in the living tissues of a host. This is termed obligate myiasis and is mostly seen in sheep, cattle, horses, and many wild animals. In people, obligate myiasis is primarily owing to the screwworm flies (Old and New World) and the human bot fly (Figs. 13.9 and 13.10). Obligate myiasis from the human bot fly of Central and South America is rarely fatal, but the condition has led to considerable pathology and death in the case of screwworm flies. Screwworm flies use livestock as primary hosts but will infest humans. If, for example, a female screwworm fly oviposits just inside the nostril of a sleeping human, hundreds of developing maggots may migrate through the turbinal mucous membranes, sinuses, and other tissues. Surgical removal of all the larvae would be extremely difficult. Fortunately, because of the sterile male release program, screwworm flies have been eliminated from the United States and Mexico.

More rarely, fly species that infest wild animals may attack humans. These cases may present as a "maggot in a boil" or other furuncular-like lesion. Since the lesion develops in otherwise healthy tissue, and since there is often no international travel history, physicians are stymied regarding the identification of these fly larvae. In one such case that this author investigated, a 3-yr-old boy somehow became infested with a bot fly larva that normally attacks squirrels, chipmunks, or rabbits [7]. The family lived in a rural area near large tracts of woods containing abundant wildlife. According to the mother, the boy complained of being "stung" by a bumble bee on his side and the neck while watching television early one morning. She said that within 5 min, typical sting-like "welts" occurred at the places the child pointed out. Within 2 d, a line of vesicles extended away from the lesions—presumably caused by the larvae migrating in the skin. The lesion on his side extended upward in a sinuous fashion about 10 cm, ending in a small papule. No further development occurred at the side lesion (apparently the larva died). The larva in his neck continued to enlarge and migrated about 4 cm laterally. After about 14 d, the dermal tumor was inflamed and contained a central opening about 3 mm in diameter, apparently through which the larva obtained air. The child often cried and complained of severe

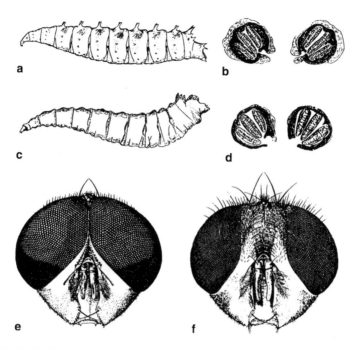

**Fig. 13.9**  Old World screwworm flies, *Chrysomya* spp. (**a**) *Chrysomya albiceps* larva, (**b**) same, showing posterior view of larval spiracles; (**c**) *Chrysomya chloropyga* larva, (**d**) same, showing posterior view of larval spiracles; (**e**) *Chrysomya megacephala*, face of male, (**f**) face of female. (From USDA publ. Ref. [5])

**Fig. 13.10**  Human bot fly larva *Dermatobia hominis*. (Photo copyright 2007 by Jerome Goddard, Ph.D.)

pain. Despite numerous trips to physicians, the myiasis was not diagnosed until almost 4 wk after the initial "stinging incident." An ER physician expressed the larva, which was ultimately forwarded to the health department for identification. On examination, the specimen was identified as a second-stage larva of a fly in the genus *Cuterebra* (the rabbit and rodent bot flies). Interestingly, adults of this type fly resemble bumble bees (Fig. 13.11).

**Fig. 13.11** Squirrel bot fly
adult, *Cuterebra.* (Photo
copyright 2014 by Jerome
Goddard, Ph.D.)

## 13.2    Contributing Factors

### 13.2.1    *Accidental Myiasis*

Accidental enteric myiasis occurs from ingesting fly eggs or young maggots on uncooked foods or previously cooked foods that have been subsequently infested. Cured meats, dried fruits, cheese, and smoked fish are commonly infested foods. I once investigated a case resulting from consumption of wild grapes (muscadines) [2]. Other cases of accidental myiasis may occur from contaminated catheters, douching syringes, or other invasive medical equipment, or sleeping with body exposed.

### 13.2.2    *Facultative Myiasis*

Several fly species lay eggs on dead animals or rotting flesh—especially blow flies and flesh flies (Fig. 13.8). Accordingly, these flies may mistakenly oviposit in a foul-smelling wound of a living animal. The developing maggots may subsequently invade healthy tissue. Facultative myiasis most often is initiated when flies oviposit in necrotic, hemorrhaging, or pus-filled lesions. Wounds with watery alkaline discharges (pH 7.1–7.5) have been reported as being especially attractive to blow flies. Facultative myiasis frequently occurs in semi-invalids who have poor (if any) medical care. Often, in the case of the very elderly, their eyesight is so weak that they do not detect the infestation. In clinical settings, facultative myiasis mostly occurs in incapacitated patients who have recently had major surgery or those having large or multiple uncovered or partially covered festering wounds. However, not all human cases of facultative myiasis occur in or near a wound. In the United States, larvae of the blowfly L. *sericata* have been reported from the ears and nose of healthy patients with no other signs of trauma in those areas [8].

## 13.2.3   Obligate Myiasis

Obligate myiasis is a zoonosis; humans are not the ordinary host but may become infested (see Textbox). Human infestation by the human bot fly is very often via a mosquito bite—the eggs are attached to mosquitoes and other biting flies; however, human screwworm fly myiasis is a result of direct egg laying onto a person, most often in or near a wound or natural orifice. Screwworm flies lay eggs during daytime.

---

**Myiasis Confused with Boils**

Boil-like lesions are often produced in cases of cutaneous myiasis; this may also be called furuncular myiasis. In furuncular myiasis, the lesion usually begins as a papule, gradually enlarging to an erythematous, dome-shaped nodule containing a central pore. Exact size of the lesion depends on the species of fly larva involved and the stage of development, but generally the nodular lesion is at least 1 cm across with an ill-defined, indurated inflammatory edema extending out about 1–2 cm. The central hole is about 3 mm in diameter and easily visible. The developing fly larva generally does not migrate through the skin (there are rare exceptions) but remains stationary, gradually increasing in size. Lesions may have a discharge containing pus, blood, and/or portions of the cast larval skin as the developing larva molts. Itching and pain accompany the infestation. The inflammatory reaction around the lesion may lead to lymphangitis and regional lymphadenopathy. Secondary infection can occur, especially if the larva dies in situ or if the patient crudely or incompletely removes the larva. Cutaneous myiasis presenting as furuncular lesions is generally not life-threatening, as opposed to myiasis caused by screwworm flies. However, considerable pain, misery, and mental anguish are associated with the infestation. And the psychological trauma should not be underestimated. I have heard several patients say things like, "Just the thoughts of that fly maggot living in my skin…" Even though furuncular myiasis does not ordinarily occur in the United States (there are a few rare exceptions), modern, rapid, international air travel has created a "global village" in which tropical maladies are easily imported.

## 13.2.4  Myiasis in Clinical Practice

Concerned patients often bring in larval specimens found in stool or in the toilet which they found in say "came out of them." Physicians and laboratory personnel must be careful not to confirm such allegations without definitive proof. Just because someone says they found a maggot in their stool does not mean it passed out of the digestive tract. Depending upon several factors, including cleanliness of the home or bathroom, the maggots may have coincidentally been found in/near stool samples or could subsequently have infested the stool samples. On a number of occasions, I have investigated cases of "maggots" in toilet bowls which were in fact soldier fly larvae coming from the wax seal (where the toilet connects to the floor). The larvae crawl away from their food source when ready to pupate and often end up in the toilet bowl.

Myiasis cases reported to the Mississippi Department of Health have included urogenital, aural, nasal, and cutaneous infestations. However, it has been my experience that most clinical samples (facultative myiasis) stem from blowfly (*Diptera*, family Calliphoridae) larvae being found in a patient's nose, ear, rectum, or pus-filled wound. If the site is a natural orifice, there is/was usually a lesion, infection, etc., which proved attractive to the female fly. Nasal lesions may be from chronic cocaine use [9]. Many times, the patient is an invalid or otherwise "exposed" and unable to care for himself. These cases of myiasis are usually not life threatening because the larvae only rarely invade healthy tissue. Patients with myiasis should be queried about recent travel history. Occasionally, human bot fly or screwworm myiasis occurs in travelers returning from tropical countries. One woman I knew personally, returning home from Belize, complained about a "boil" behind her ear. She claimed she could hear a "clicking" sound inside the boil. Eventually she was seen by a physician who diagnosed human bot fly myiasis. Apparently, she really was hearing the larva as it moved or fed inside the tissues near her ear.

## 13.2.5  Differential Diagnosis

Boil-like and/or nodular lesions on the human skin can have numerous causes, including staphylococcal infections, cat-scratch disease, tick-bite granuloma, tungiasis (infestation by a burrowing flea), and infestation with various parasitic worms (such as *Dirofilaria*, *Loa loa*, and *Onchocerca*), as well as many other causes. Many nodular lesions eventually ulcerate if the inflammatory process is intense enough to result in destruction of the overlying epidermis. Lesions from myiasis do not ulcerate. The central core of the lesion should be examined for evidence of a fly larva. Sometimes the posterior end of the larva is clearly visible just below the skin surface. Another helpful clue in diagnosing myiasis with the human bot fly, *Dermatobia hominis*, is that sometimes the pointed posterior end of the larva protrudes from the central opening. This protrusion may be visible on several occasions and extend up to 5 mm above the skin.

## 13.3  Prevention, Treatment, and Control

Prevention and sanitation can avert much accidental and facultative myiasis occurring in the industrialized world. Exposed foodstuffs should not be unattended for any length of time to prevent flies from ovipositing therein. Covering, and preferably refrigerating, leftovers should be done immediately after meals. Washing fruits and vegetables prior to consumption can help remove developing maggots, although visual examination should also be accomplished during slicing or preparing these items. Other forms of accidental myiasis may be prevented by protecting invasive medical equipment from flies and avoiding sleeping nude, especially during daytime. To prevent facultative myiasis, extra care should be taken to keep wounds clean and covered, especially on elderly or helpless individuals. Daily or weekly visits by a home health nurse can help prevent facultative myiasis in patients who stay at home. In institutions containing invalids, every effort should be made to control entry of flies into the facility. This might involve such things as keeping doors and windows screened and in good repair, thoroughly sealing all cracks and crevices, installing air curtains over doors used for loading and unloading supplies, and installing UV fly traps in areas accessible to the flies but inaccessible to patients. Prevention of obligate myiasis involves avoiding sleeping outdoors during daytime in screwworm-infested areas and using insect repellents in Central and South America to prevent bites by bot fly egg-bearing mosquitoes.

Treatment of accidental enteric myiasis is probably not necessary (although there may be rare instances of clinical symptoms), since in most cases there is no development of the fly larvae within the highly acidic stomach environment and other parts of the digestive tract. They are killed and merely carried along through the digestive tract. Treatment of other forms of accidental myiasis as well as facultative or obligate myiasis involves removal of the larvae. Alexander [10] recommends debridement with irrigation. Others have suggested surgical exploration and removal of fly larvae under local anesthesia [8]. Care should be taken not to burst the maggots on removal. Human bot fly larvae have been successfully removed using "bacon therapy," a treatment method involving covering the punctum (breathing hole in the patient's skin) with raw meat or pork [11]. In a few hours, the larvae migrate into the meat and are then easily extracted. Maggot infestation of the nose, eyes, ears, and other areas may require surgery if larvae cannot be removed via natural orifices. Since blow flies and other myiasis-causing flies lay eggs in batches, there could be tens or even hundreds of maggots in a wound.

# References

1. Mazzotti L. Casos humanos de miasis intestinal. Ciencia, Mex. 1967;5:167–168.
2. Goddard J, Hoppens K, Lynn K. Case report on enteric myiasis in Mississippi. J Miss State Med Assoc. 2014;55(4):132–3.
3. Greenberg B. Two cases of human myiasis caused by *Phaenicia sericata* in Chicago area hospitals. J Med Entomol. 1984;21:615.
4. Merritt RW. A severe case of human cutaneous myiasis caused by *Phaenicia sericata*. Calif Vector Views. 1969;16:24–6.
5. USDA. Pests infesting food products. United States Department of Agriculture, ARS, Agri. Hndbk. No. 655, p. 213 ; 1991.
6. Richard RD, Ahrens EH. New distribution record for the recently introduced blow fly *Chrysomya rufifaces* in North America. Southwest Entomol. 1983;8:216–8.
7. Goddard J. Human infestation with rodent bot fly larvae: a new route of entry? South Med J. 1997;90:254–5.
8. Anderson JF, Magnarelli LA. Hospital acquired myiasis. Asepsis. 1984;6:15.
9. Goddard J, Varnado WC. Human nasal myiasis possibly associated with cocaine abuse. J Entomol Sci. 2010;45:188–9.
10. Alexander JO. Arthropods and human skin. Berlin: Springer-Verlag; 1984. p. 422.
11. Brewer TF, Wilson ME, Gonzalez E, Felsenstein D. Bacon therapy and furuncular myiasis. J Am Med Assoc. 1993;270:2087–8.

# Chapter 14
# Imaginary Insect or Mite Infestations

## 14.1 Introduction and Medical Significance

If in practice very long, most physicians, regardless of specialty, have encountered patients who claim that invisible insects or mites are on/in their skin. For proof, they may even bring in tiny bottles, bags, envelopes, and so forth, containing specks of dusts, hair, lint, or skin that they claim contain the offending specimens. In response, these patients are usually examined for actual arthropod infestations, evaluated for organic causes of the crawling sensations, and (frequently) given antiscabicidal creams or lotions. However, more often than not, the patient becomes discouraged with that particular doctor and moves on to another. Such wandering among physicians, entomologists, and public health personnel may last for years without the patient ever receiving the help he or she really needs.

This condition, often called delusions of parasitosis (DOP), is a psychiatric disorder characterized by an unshakable belief that tiny, almost invisible insects or mites are living on or in the body. No argument or scientific evidence can convince a patient with true DOP that there is no infestation [1, 2]. A condition consistent with DOP was first recognized by Thibierge [3] in the late 1800s, but appropriate definition and terminology were not applied until later. It has been called Ekbom's syndrome, delusionary parasitosis, delusory parasitosis, and others. Wilson and Miller designated the condition "delusions of parasitosis," which seems to be accurate and a term widely used [4]. However, a few people refer to the condition as psychogenic parasitosis based on a study in which many DOP patients gave up the belief (that bugs were on them) after reassurance and suggestion [5]. The authors concluded that a delusion is a fixed false belief by definition, and therefore, any patients who had a shakable belief could not be considered delusional in the classic sense [5]. Regardless of the naming controversy, adverse health effects from DOP include radical patient efforts to rid themselves of the "bugs," such as quitting jobs, burning furniture, abandoning homes, and using powerful pesticides dangerously. Sometimes patients commit suicide [6].

© Springer International Publishing AG, part of Springer Nature 2018
J. Goddard, *Infectious Diseases and Arthropods*, Infectious Disease,
https://doi.org/10.1007/978-3-319-75874-9_14

One man I knew piled all his household furniture in the backyard and burned it. His comment at that time was "the house is next if this doesn't get 'em."

## 14.2   Clinical Aspects and Contributing Factors

The patient is characteristically an elderly female [7, 8]. It has been my experience that younger patients (<50) are usually male [9] and, if in their 20s, associated with methamphetamine or cocaine abuse. Most patients present with complaints of tiny insects or mites crawling under their skin, biting, tickling, or burrowing. Seldom is itching the primary complaint. Lesions may be present, though neurotic excoriation may be the cause [10]. Other skin damage may be present resulting from intense scrubbing (steel wool, metal scratch pads, and so on) or use of harsh chemicals, such as gasoline or Clorox. In one study, 82% of DOP patients presented with "evidence" of their infestation that included tiny, nonharmful insects, dust, specks of debris, and skin or ear scrapings wrapped in paper or in jars or vials [9] (Fig. 14.1). A consistent and diagnostic feature is the patient's absolute conviction that he or she knows exactly what is going on. The patient may also be angry that his or her physician cannot even see, much less eliminate, the "bugs." The medical history often has a persuasive, yet idiosyncratic logic, and the patient may be so convincing that others in the family secondarily share in the delusion—a *folie a deux*.

Various events, such as sudden family bereavement, flooding, or exposure to parasitized persons or animals, have been cited as precipitating factors [7]. Abuse of drugs such as methamphetamine, phencyclidine (PCP), and cocaine may lead to

**Fig. 14.1**  Samples in folded pieces of paper sent in by a DOP patient

DOP [11]—one case in the literature was clearly attributed to cocaine use [12]. Sometimes an initial and real insect infestation in the home environment triggers the delusion. For example, if someone with an indoor pet gets fleas inside the home, he or she may still feel mysterious biting long after the fleas have been killed by an exterminator. The recent advent of bed bug infestations has exacerbated the DOP problem, both by people actually (one time) having bed bugs or just being obsessively fearful they will get bed bugs [13].

## 14.3 Morgellons Disease

The term, *Morgellons disease,* referring to a medical case described in 1674, is often interchangeably used with DOP. Patients who say they have Morgellons have an unshakable belief that fibers are emerging from their skin and that the fibers are causing their biting and stinging sensations. Along with a wide range of skin conditions, people with this disorder also often exhibit fatigue, arthralgia, cognitive decline, and mood disorders. In fact, one study showed that 70% of 115 Morgellons patients complained of chronic fatigue [14]. Morgellons may be psychologically "infective," leading to other family members and coworkers (seemingly) getting the disease. At this time, there is no evidence incriminating insects or mites as the cause of Morgellons. Many, if not most, physicians do not recognize Morgellons as a valid clinical entity and generally view the disorder in three ways [15]:

1. Morgellons disease is a real and specific condition requiring confirmation by future research.
2. The signs and symptoms of Morgellons disease are caused by other conditions, often mental illness.
3. Maybe we should reserve judgment until more information is available.

## 14.4 Differential Diagnosis

DOP falls under the Diagnostic and Statistical Manual of Mental Disorders, Fifth Edition (DSM-5) in the category of delusional disorder somatic type [16]. To meet the diagnostic criteria, the "infestation" should last at least a month and not be part of any schizophrenia manifestations [16]. DOP must be separated from actual insect or mite infestations, as well as from chemical or organic conditions contributing to a crawling or tingling sensation on the skin. Prescription drugs are often a cause of such sensations [2]. Bhatia et al. [8] provided an excellent clinical profile of 52 DOP cases which is helpful to clinicians for diagnosis. Skin scrapings by a dermatologist may be indicated to rule out scabies. Samples submitted by the patient should be examined for the presence of biting insects or mites. Sometimes lab personnel can accomplish this, although a local university entomology department or Extension

Service may be the better alternative. Ideally, the patient's home should be inspected for biting arthropods. Pest controllers will perform this service for a fee but may prey on patient fears and recommend expensive pesticidal treatments. Health department or university personnel sometimes become involved in home visits but are under no mandate to investigate private pest problems.

There may be internal physiological causes of the crawling sensation. Diabetes, icterus, atopic dermatitis, and lymphoblastomas have skin manifestations that can mistakenly be considered arthropod-induced [17, 18]. At times, pellagra may produce DOP, which disappears with appropriate therapy [18].

## 14.5   Treatment Strategies

An interdisciplinary approach is needed to help DOP patients, mainly involving family practice physicians, dermatologists, psychiatrists, and entomologists [19] (Fig. 14.2). Family practice physicians or dermatologists are usually the providers who first see DOP patients. Physicians need to be careful not to diagnose "insect bites" based on lesions alone and should be careful not to accept or feed into the patient's delusions by giving the impression that the delusion is believed to be real [16]. In addition, physicians should call upon entomologists or public health laboratories to examine samples. Entomologists need to understand the medical complexity of delusions—that there are intensive obsessional worries, true delusions, and a whole host of abnormal personality traits associated with DOP—and avoid any hint of medical evaluation of the patient. Although psychiatric evaluation is needed, most DOP patients will not see a psychiatrist (even if referred). Instead,

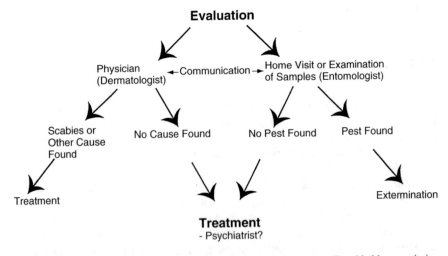

**Fig. 14.2** Possible strategy for assessing patient with mysterious bites. (Provided by permission from Ref. [15])

they will seek out another physician, thus starting the whole process over again. For this reason, Koblenzer, a dermatologist, says "because the patient has great emotional involvement in the skin, I usually allow him or her to maintain that focus, but I substitute positive healing measures for their prior destructive rituals. Supervision of topical treatments through frequent, even quite short office visits, serves to allow a supportive and accepting relationship to develop. Hopefully, this will gradually allow the patient to accept either oral medication or referral to a psychiatrist." [20]. Traditionally, one of the most extensively used drugs for DOP has been the first-generation antipsychotic agent, pimozide, although other medications, such as haloperidol, risperidone, and olanzapine, have also been used with success [16, 18]. Controlled studies (although with few patients) have shown a response rate of 54–90% to pimozide [18, 21]. Pimozide has several serious effects and should only be used with careful supervision. The most common side effects are parkinsonian symptoms, such as tremor, bradykinesia, shuffling gait, and masked facies. Tardive dyskinesia is perhaps the most worrisome, since it may be irreversible. If pimozide is used, the lowest effective dosage should be used for the shortest possible duration, because many patients with DOP fit the profile for the patient at highest risk for tardive dyskinesia (a woman >50 years old) [21].

# References

1. Koblenzer CS. The clinical presentation: diagnosis and treatment of delusions of parasitosis - a dermatologic perspective. Bull Soc Vector Ecol. 1993;18:6–10.
2. Hinkle N. Ekbom syndrome: the challenge of "invisible bug" infestations. Annu Rev Entomol. 2010;55:77–94.
3. Thibierge G. Les acarophobes. Rev Gen Clin Ther. 1894;32:373–6.
4. Wilson JW, Miller HE. Delusion of parasitosis. Arch Dermatol Syphilol. 1946;54:39–56.
5. Zanol K, Slaughter J, Hall R. An approach to the treatment of psychogenic parasitosis. Int J Dermatol. 1998;37:56–63.
6. Monk BE, Rao YJ. Delusions of parasitosis with fatal outcome. Clin Exp Dermatol. 1994;19:341–2.
7. Alexander JO. Arthropods and human skin. Berlin: Springer; 1984. p. 422.
8. Bhatia MS, Jagawat T, Choudhary S. Delusional parasitosis: a clinical profile. Int J Psychiatry Med. 2000;30:83–91.
9. Goddard J. Analysis of 11 cases of delusions of parasitosis reported to the Mississippi Department of Health. South Med J. 1995;88:837–9.
10. Obermayer ME. Dynamics and management of self-induced eruptions. Calif Med. 1961;94:61–71.
11. Trabert W. 100 years of delusional parasitosis: meta-analysis of 1,223 case reports. Psychopathology. 1995;28:238–46.
12. Elpern DJ. Cocaine abuse and delusions of parasitosis. Cutis. 1988;42:273–4.
13. Rieder E, Hamalian G, Maloy K, Streicker E, Sjulson L, Ying P. Psychiatric consequences of actual versus feared and perceived bed bug infestations: a case series examining a current epidemic. Psychosomatics. 2012;53:85–91.
14. Pearson ML, Selby JV, Katz KA, Cantrell V, Braden CR, Parise ME, et al. Clinical, epidemiologic, histopathologic and molecular features of an unexplained dermopathy. PLoS One. 2011;7(1):e29908.

15. Anonymous. Morgellons disease. Mayo Clinic health information factsheet, www.mayoclinic. com/health/morgellons-disease/sn00043; 2011.
16. Mostaghimi L. Psychocutaneous medicine. In: Bope ET, Kellerman RD, editors. Conn's current therapy. Philadelphia: Elsevier; 2017. p. 963–8.
17. Goddard J. Physician's guide to arthropods of medical importance. 6th ed. Boca Raton: Taylor and Francis (CRC); 2013. p. 412.
18. Suh KN, Keystone JS. Delusional parasitosis. In: Guerrant RL, Walker DH, Weller PF, editors. Tropical infectious diseases: principles, pathogens, and Pratcice, vol. 2. 2nd ed. Philadelphia: Churchill Livingstone; 2006. p. 1700–7.
19. Goddard J. Imaginary insect or mite infestations. Inf Med. 1998;15:168–70.
20. Koblenzer CS. Psychocutaneous disease. Orlando: Grune and Stratton; 1987. p. 328.
21. Driscoll MS, Rothe MJ, Grant-Kels JM, Hale MS. Delusional parasitosis: a dermatologic, psychiatric, and pharmacologic approach. J Am Acad Dermatol. 1993;29:1023–33.

# Appendix 1: Signs and Symptoms of Arthropod-Borne Diseases

The following is an alphabetical listing of common signs and symptoms of arthropod-borne diseases. Unfortunately, few signs and symptoms are specific to any one disease. Further differentiation by appropriate laboratory or radiologic tests may be needed. By no means should this listing be considered as a complete differential diagnosis of any of the symptoms discussed.

**Adenopathy**  Generalized adenopathy may occur in the early stages of African trypanosomiasis—the glands of the posterior cervical triangle being most conspicuously affected (Winterbottom's sign). Adenopathy may also be seen in the acute stage of Chagas' disease.

**Anemia**  Anemia may be seen in cases of malaria, babesiosis, and trypanosomiasis. Anemia can be especially severe in falciparum malaria. It may also occur in people (especially the elderly) who have large populations of bed bugs in their homes.

**Blister**  A blister may occur at arthropod bite sites. Blistering may also occur as a result from blister beetles contacting the human skin.

**Bull's-Eye Rash (*see* Erythema Migrans)**

**Chagoma**  An indurated, erythematous lesion may occur on the body—often the head or neck—caused by *Trypanosoma cruzi* infection (Chagas'disease). A chagoma may persist for 2–3 mo.

**Chyluria**  The presence of chyle (lymphatic fluid) in the urine is often seen in lymphatic filariasis. Urine may be milky white and even contain microfilariae.

© Springer International Publishing AG, part of Springer Nature 2018
J. Goddard, *Infectious Diseases and Arthropods*, Infectious Disease,
https://doi.org/10.1007/978-3-319-75874-9

**Coma**  Sudden coma in a person returning from a malarious area may indicate cerebral malaria. African trypanosomiasis (sleeping sickness) may also lead to coma after a long period of increasingly severe symptoms of meningoencephalitis. Rocky Mountain spotted fever and other rickettsial infections may also lead to coma.

**Conjunctivitis**  Chagas' disease and onchocerciasis may lead to chronic conjunctivitis. Also, Zika virus infection may cause conjunctivitis.

**Dermatitis**  Several arthropods may directly or indirectly cause dermatitis. Chiggers and other mites may attack the skin, causing a maculopapular rash. Scabies mites may burrow under the skin's surface making itchy trails or papules. Lice may give rise to hypersensitivity reactions with itchy papules. Chigoe fleas burrow in the skin (especially on the feet), causing local irritation and itching. Macules or erythematous nodules may result as a secondary cutaneous manifestation of leishmaniasis.

**Diarrhea**  Leishmaniasis (and specifically visceral leishmaniasis—kala-azar) may lead to mucosal ulceration and diarrhea. In falciparum malaria, plugging of mucosal capillaries with parasitized red blood cells may lead to watery diarrhea.

**Edema**  Edema may result from arthropod bites or stings. Loiasis (a nematode worm transmitted by deer flies) may also cause edema—a unilateral circumorbital edema as the adult worm passes across the eyeball or lid. Passage of the worm is brief, but inflammatory changes in the eye may last for days. Loiasis may also lead to temporary appearance of large swellings on the limbs, known as Calabar swellings at the sites where migrating adult worms occur. Unilateral edema of the eyelid, called Romaña's sign, may occur in Chagas' disease. African trypanosomiasis (sleeping sickness) may result in edema of the hips, legs, hands and face.

**Elephantiasis**  Hypertrophy and thickening of tissues, leading to an "elephant leg" appearance, may result from lymphatic filariasis. Various tissues may be affected, including limbs, the scrotum, and the vulva.

**Eosinophilia**  Helminth worms may cause eosinophilia. Atopic diseases, such as rhinitis, asthma, and hay fever, also are characterized by eosinophilia.

**Eosinophilic Cerebrospinal Fluid Pleocytosis**  Cerebrospinal fluid eosinophilic pleocytosis can be caused by a number of infectious diseases (including rickettsial and viral infections) but is primarily associated with parasitic infections.

**Epididymitis**  Epididymitis, with orchitis, may be an early complication of lymphatic filariasis.

**Erythema Migrans**  Erythema migrans may follow bites of ticks infected with the causative agent of Lyme disease, *Borrelia burgdorferi*. Typically the lesion consists of an annular erythema with a central clearing surrounded by a red migrating border. Although erythema migrans does not always occur, it can be highly diagnostic for Lyme disease. Also, erythema migrans-like lesions may occur after tick and other arthropod bites as a result of a hypersensitivity reaction to their saliva.

**Eschar** A round (generally 5–15 mm) spot of necrosis may result from boutonneuse fevers, American boutonneuse fever, (spotted fever group illnesses), or scrub typhus. An eschar develops at the site of tick or chigger bite.

**Excoriation** Lesions produced by "self-scratching" may be a sign of imaginary insect or mite infestations (delusions of parasitosis).

**Fever** Fever is a common sign of many arthropod-borne diseases, including the rickettsioses, typhus, dengue, yellow fever, plague, the encephalitides, and others. In some cases, there are cyclical peaks of fever, such as in relapsing fever (tick-borne) or malaria. Falciparum malaria is notorious for causing extremely high fever (107 °F or higher). Filariasis may be marked by fever, especially early in the course of infection.

**Hematemesis** Coffee-ground color or black vomit may be a sign of yellow fever.

**Hemoglobinuria** Falciparum malaria can cause "blackwater fever."

**Hydrocele** Hydrocele may result from lymphatic filariasis, developing as a sequel to repeated attacks of orchitis.

**Keratitis** Inflammation of the cornea is sometimes a result of ocular migration of *Onchocerca volvulus* microfilariae. It may lead to blindness.

**Leukopenia** Leukopenia is a prominent finding in cases of ehrlichiosis. It may also occur (3000–6000/mm3) with a relative monocytosis during the afebrile periods of malaria.

**Lymphadenitis** Inflammation of one or more lymph nodes may be a sign of lymphatic filariasis—especially involving the femoral, inguinal, axillary, or epitrochlear nodes.

**Lymphangitis** Lymphangitis can be an early symptom of lymphatic filariasis, involving the limbs, breast, or scrotum.

**Lymphocytosis** Lymphocytosis may occur in Chagas' disease.

**Maggots** The presence of fly larvae in human tissues is termed myiasis. Various blow flies, bot flies, and other muscoid flies are usually involved.

**Meningoencephalitis** Meningoencephalitis has many causes but may be a result of trypanosomes in the case of African trypanosomiasis (sleeping sickness) or

Chagas' disease (although generally milder). Falciparum malaria infection may be cerebral, with increasing headache and drowsiness over several days, or even sudden onset of coma. Mosquito-borne encephalitis viruses such as West Nile may also cause meningoencephalitis.

**Myocarditis** Chagas' disease may lead to myocardial infection. African trypanosomiasis may also cause myocarditis to a lesser extent.

**Neuritis** Neuritis may be caused by bee, ant, or wasp venom. Occasionally stings to an extremity result in weakness, numbness, tingling, and prickling sensations for days or weeks. Neuritis may also result from infection with the Lyme disease spirochete.

**Nodules, Subcutaneous** Onchocerciasis may present as skin nodules (*see* Onchocercoma). Tick bites may also result in nodules. Fly larvae in the skin (myiasis) may also present as nodules. Common species involved are the human bot fly larva, *Dermatobia hominis*; the tumbu fly, *Cordylobia anthropophaga*; and rodent bot fly larvae, *Cuterebra spp.*

**Onchocercoma** Coiled masses of adult *O. volvulus* worms beneath the skin enclosed by fibrous tissues may occur in patients living in tropical countries endemic for onchocerciasis.

**Orchitis** Orchitis may be a symptom of lymphatic filariasis; repeated attacks may lead to hydrocele.

**Paralysis** Ascending flaccid paralysis may result from tick attachment. The paralysis is believed to be caused by a salivary toxin injected as the tick feeds.

**Proteinuria** Proteinuria, with hyaline and granular casts in the urine, often occurs in falciparum malaria.

**Puncta** A small, point-like pierce mark may mark the bite or sting site of an arthropod. Paired puncta may indicate spider bite or centipede bite.

**Rash** There are myriad causes of rash, but rash may accompany many arthropod-borne diseases, such as Rocky Mountain spotted fever, ehrlichiosis, murine typhus, and African trypanosomiasis. The rash may appear to be ring-like and expanding in the case of Lyme disease (*see* **Erythema Migrans**). An allergic urticarial rash may be seen in the case of bites or stings.

**Romaña's Sign** A common sign early in the course of Chagas' disease, Romaña's sign is a unilateral palpebral edema, involving both the upper and lower eyelids. This generally occurs when the Chagas' organism enters via the eye, usually by rubbing.

**Shock**  Shock may occur from arthropod stings (rarely bites) as a result of hypersensitivity reactions to venom or saliva. Shock may also accompany falciparum malaria.

**Splenomegaly**  Splenomegaly can be a result of lymphoid hyperplasia in both African and American trypanosomiasis. It may also occur in visceral leishmaniasis (kala-azar).

**Tachycardia**  Both African and American trypanosomiasis may produce tachycardia. In Chagas' disease tachycardia may persist into the chronic stage where it may be associated with heart block.

**Ulcers, Cutaneous**  A shallow ulcer (slow to heal) may be a sign of cutaneous leishmaniasis. In the New World, lesions from cutaneous leishmaniasis are often found on the ear. Also, a firm, tender, raised lesion up to 2 cm or more in diameter may occur at the site of infection in African trypanosomiasis.

**Urticaria**  Urticaria may result from an allergic or generalized systemic reaction to arthropod venom or (more rarely) saliva.

**Verruga Peruana**  A benign dermal eruption (Peruvian warts) is one manifestation of bartonellosis. The verrugae are chronic, lasting from several months to years, and contain large numbers of *Bartonella bacilliformis* bacteria.

**Winterbottom's Sign**  In the early stages of African trypanosomiasis, patients may exhibit posterior cervical lymphadenitis.

# Appendix 2: Diagnostic Tests Used in Arthropod-Borne Diseases

## Agglutination

Agglutinations are antibodies that cause clumping together (agglutination) of microorganisms, erythrocytes, and often antigenic particulates. If the serum being tested is specific, agglutinins present will cause cultured parasites or bacteria to clump when the serum is introduced.

## Complement Fixation

In CF tests, the suspected serum is incubated with a known source of antigen, permitting the antigen-antibody interaction to bind complement and remove it from the reaction mixture. A sheep-blood indicator is then added which hemolyzes in the presence of free complement. If the sheep cells fail to hemolyze, complement is absent; its absence testifies to the prior occurrence of an antigen-antibody reaction. By varying the serum or antigen dilution, one can achieve a crude approximation of titer.

## Direct Fluorescent Antibody

A DFA test (some texts refer to it as direct immunofluorescence or DIF) utilizes fluorescent tagging of antibodies produced against the pathogen in question. These tagged antibodies can be purchased commercially against a wide variety of organisms. When tagged antibodies are placed on a microscope slide containing the pathogen, the organisms fluoresce when viewed by fluorescent microscopy. DFA is a one-step procedure involving the placement of tagged antibody on a suspect smear of tissue or blood and viewing (after a brief phosphate-buffered saline [PBS] wash) with a UV light-equipped microscope.

© Springer International Publishing AG, part of Springer Nature 2018
J. Goddard, *Infectious Diseases and Arthropods*, Infectious Disease,
https://doi.org/10.1007/978-3-319-75874-9

## Enzyme-Linked Immunosorbent Assay (ELISA)

Similar, if not identical, to a test called enzyme immunoassay (EIA), the ELISA test may be used for quantitative determination of either antigen or antibody. The appropriate antigen or antibody is bound to (usually) plastic microtiter plates, and the specimen to be tested is then added and given time to react with the already present antigen or antibody. After a wash to remove any unbound test material, an enzyme-linked antigen or antibody is added. After a second wash, a substrate is added that will react with the remaining enzyme to produce a color change.

## Hemagglutination Inhibition (HI)

The HI test measures the presence of hemagglutination-inhibiting antibody toward a particular organism. The suspected serum is incubated with fluid medium known to be capable of agglutinating red cells. After the incubation period, the agglutinating potency is measured, and the absence of subsequent agglutination indicates the presence of specific antibodies in the serum.

## Immunohistochemistry (IHC)

IHC is used to visualize pathogens in tissues as well as to diagnose abnormal cells such as those found in cancerous tumors. The test is performed on tissue sections and, in most cases, utilizes an antibody conjugated to an enzyme, such as peroxidase, that can catalyze a color-producing reaction. Alternatively, the antibody can be tagged to a fluorescent chemical such as FITC, rhodamine, or Texas Red, for reading with a fluorescent microscope.

## Indirect Fluorescent Antibody (IFA)

The IFA test is a two-step test involving the placement of patient serum suspected of containing antibodies on a slide with fixed, known antigen. After an incubation period and PBS washing, the slide is then covered with a solution containing fluorescent-tagged antihuman antibodies. After a second incubation period and PBS washing, the slide is viewed by fluorescent microscopy. Fluorescence of antigen on the slide is considered evidence of patient antibodies toward that particular organism. By serially diluting patient serum, a titer can be determined.

## Leishmanin (Montenegro Test)

The leishmanin test (not available in the United States) has historically been used to help diagnose cases of cutaneous and mucocutaneous leishmaniasis. It involves an intradermal injection of a suspension of killed promastigotes. A high percentage of *Leishmania tropica* and *Leishmania braziliensis* infections will test positive by this test.

## Mazzotti

The Mazzotti test is used to determine if a patient has onchocerciasis. It can be dangerous and is only rarely used. It consists of oral administration of 25 or 50 mg of diethylcarbamazine to a patient suspected of having onchocerciasis. If the patient is infected, an intense itching occurs in a few hours (as the microfilariae die within the skin). The itching is then controlled by short-term administration of corticosteroids or will subside on its own within 2–3 d.

## Neutralization

The neutralization test (NT) is the most specific immunologic test for the majority of viral infections. The identification of an unknown viral isolate is made by analyzing the degree to which antisera of known reactivity prevent the virus from infecting tissue-culture cells, eggs, or animals. If neutralizing antibody is present, virus cannot attach to cells, and infectivity is blocked.

## Polymerase Chain Reaction (PCR)

PCR has dramatically changed diagnostic microbiology in recent years. PCR makes specific identification of pathogens possible, even when only a few organisms are present. PCR is a highly sensitive technique by which minute quantities of DNA or RNA sequences are enzymatically amplified to the extent that a sufficient quantity of material is available to reach a threshold signal for detection using a specific probe. The scientific basis of PCR is that each infectious disease agent (in fact, every living thing) possesses a unique signature sequence in its DNA or RNA by which it can be identified. In other words, there is a unique sequence of amino acids for each organism. By finding those unique sequences and constructing primers to amplify those specific areas of DNA, identification of an organism can be accomplished from a blood or tissue sample or even from an infected arthropod vector. PCR is carried out using a thermocycler, which produces a series of heat–cool cycles, whereby double-stranded DNA is dissociated into single strands that are in turn allowed to anneal in the presence of specific primers on cooling.

Through the successive heat–cool cycles (usually about 30), the DNA sequence to be detected is amplified millions of times. The product is then visualized after separation on agarose gels by electrophoresis and appropriate staining. There are various types of PCR, such as real-time PCR which allows more samples to be processed at once, and nested PCR which is more sensitive than either real-time or direct PCR.

# Appendix 3: Personal Protection from Arthropods

## Avoidance and Protective Clothing

There are some practical, nonchemical measures to prevent arthropod exposure include limiting outdoor activity and avoiding known arthropod-infested areas during the peak season. For example, people trying to avoid mosquito bites can limit outdoor activity after dark, and people in blackfly-infested areas may try to avoid outdoor activity during peak season (spring and early summer months). If outdoor activity is required, people can wear long sleeves and long pants to physically prevent biting (Fig. A.3.1). Wearing rubber boots and placing pant legs inside them is very effective against ticks, chiggers, and fire ants, especially when repellent is applied along the upper rim of the boots (Fig. A.3.2).

## Screening of Windows and Doors

One of the most basic and effective sanitation measures to limit arthropod–human contact is that of screening windows and doors. Screens can be made of metal or plastic and are ordinarily 16 × 16 × 20 mesh. They should be tight fitting over window openings. Screen doors should be hung so that they open outward. Being such a low-tech protection technique, screens are sometimes overlooked; however, their importance cannot be overemphasized. I personally investigated a death from eastern equine encephalitis in which the family had no screens on the house. When asked, the family members stated, "Mosquitoes eat us up every night!"

© Springer International Publishing AG, part of Springer Nature 2018          261
J. Goddard, *Infectious Diseases and Arthropods*, Infectious Disease,
https://doi.org/10.1007/978-3-319-75874-9

**Fig. A.3.1** Long sleeves provide effective protection against mosquito bites. (CDC photo by Lauren Bishop)

**Fig. A.3.2** Tick repellent being sprayed on boots. (Photo by Jerome Goddard, Ph.D.)

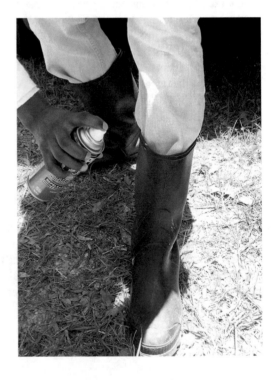

**Fig. A.3.3** Head net for protection against mosquitoes and other biting flies. (Photo courtesy Joseph Goddard)

# Netting

During mosquito season, people may choose to wear protective head gear or jackets made of netting when outdoors (Fig. A.3.3). Also, those camping can sleep under mosquito nets for protection from mosquitoes. These protection measures become mandatory on safaris or other trips to the tropics or subtropics for protection against disease-carrying mosquitoes. Ordinarily, mosquito netting is made of cotton or nylon with 23 to 26 meshes per inch. Netting should not be allowed to contact the head or body, because mosquitoes can feed through the net where it touches the skin; therefore, a frame can be used to keep the net away from the body. Insecticide-treated nets (ITNs) (Fig. A.3.4) are a very effective part of the worldwide strategy against malaria in Africa, Southeast Asia, and South America, and they are widely promoted by the World Health Organization and charitable groups such as the Bill and Melinda Gates Foundation. In a study in New Guinea, bites from anopheline mosquitoes ranged from 6.4 to 61.3 bites per person per day before distribution of bed nets and from 1.1 to 9.4 bites for 11 months after distribution [1]. ITNs are usually treated with pyrethroids such as permethrin or deltamethrin and may remain effective for months or even a year or two if not rinsed or washed; however, insecticide resistance to the pyrethroids used in ITNs has been developing, leading to major loss of efficacy [2].

**Fig. A.3.4** Bed net (rolled up above bed in this case) for protection against malaria mosquitoes. (CDC photo by Dr. B. K. Kapella)

## Repellents

There is a wide variety of insect repellents available to the public, but the most common ones contain the active ingredients DEET, picaridin, or various plant-based substances such as oil of lemon eucalyptus (Fig. A.3.5). Previously called *N,N*-diethyl-*m*-toluamide, DEET remains the gold standard of insect repellents. The chemical, discovered by US Department of Agriculture (USDA) scientists, was registered for use by the public in 1957. Twenty years of empirical testing of more than 20,000 other chemical compounds have never produced any product with the duration of protection and broad-spectrum effectiveness of DEET [3]. DEET is sold under numerous brand names and is formulated in various ways and concentrations—creams, lotions, sprays, extended-release formulations, etc. Concentrations of DEET range from about 5% to 100%, and, generally, products with higher concentrations of DEET have longer repellence times [4]; however, the correlation between concentration and repellency breaks down at some point. For example, in one study, 50% DEET provided 4 hours of protection against *Aedes aegypti* mosquitoes, but increasing the concentration to 100% provided only 1 additional hour of protection [5]. As for negative effects, DEET is absorbed through the skin into the systemic circulation, and this rarely may lead to toxic encephalopathy [6]. One study showed that about 10–15% of each dose can be recovered from urine. Other studies have shown lower skin absorption values in the range of 5.6–8.4%. Regardless, the lowest concentration of DEET providing the longest repellency should be chosen. Generally, 10–35% DEET will provide adequate protection from

**Fig. A.3.5** Some common insect repellents. (Photo by Jerome Goddard, Ph.D.)

most biting insects. Children and infants >2 months old should not be exposed to concentrations higher than 30% [6].

DEET is also effective against ticks. One study [7] demonstrated that DEET on military uniforms provided between 10% and 87.5% protection against ticks, depending on species and life stage of the tick. Overall, there was 59.8% protection against all species of ticks. Protection levels in the 50% range are less than desirable (especially for disease prevention). In a US Army repellent rating system, DEET is assigned a 2× value, whereas permethrin is given a 3× rating [8]. DEET products are simply not as effective in protecting from tick infestation as permethrin products (see discussion of permethrin below); however, the advantage of DEET is that it can be applied to the human skin.

Picaridin, sometimes known as Bayrepel®, is an alternative to DEET products that provides long-lasting protection against mosquito bites [9] (Fig. A.3.6). This relatively new repellent has been used worldwide since 1998 and is available in concentrations of 5–20%. Compared to DEET, picaridin is nearly odorless, does not cause skin irritation, and has no adverse effect on plastics; however, it does not always provide protection for as long as DEET [10, 11]. For example, a field study demonstrated 5-h protection against *Culex annulirostris* mosquitoes with picaridin vs. 7-h protection with DEET [10].

There are a variety of plant-derived substances that may provide repellency against mosquitoes including citronella, cedar, verbena, lemon eucalyptus, pennyroyal, geranium, lavender, pine, cajeput, cinnamon, rosemary, basil, thyme, allspice,

**Fig. A.3.6** Picaridin insect
repellent. (Photo by
Jerome Goddard, Ph.D.)

garlic, and peppermint. Many commercially available products with these active ingredients provide temporary protection from mosquitoes but some, almost none at all [12]. One study testing DEET-based products against Buzz Away® (containing citronella, cedarwood, eucalyptus, lemongrass, alcohol, and water) and Green Ban® (containing citronella, cajuput, lavender, safrole-free sassafras, peppermint, bergaptene-free bergamot, calendula, soy, and tea tree oils) demonstrated no repellency against *Aedes aegypti* [4]. Other studies with Buzz Away®, however, have shown repellency for about 2 h [13]. An organic repellent that was released in the United States in 1997, Bite Blocker® (containing soybean oil, geranium oil, and coconut oil), has demonstrated fairly good repellency against mosquitoes [12]. In addition, oil of lemon eucalyptus (OLE), *p*-methane 3,8-diol, has performed well in a number of recent scientific studies and is now listed on the CDC website as an alternative to DEET. In field studies against malaria-transmitting mosquitoes, OLE provided up to 6 h of protection against mosquito bites [6].

Permethrin, a pesticide that is sometimes used as a repellent, is a synthetic pyrethroid available for use against mosquitoes and ticks, but it can only be used on clothing (Fig. A.3.7). The product is sold in lawn, garden, or sporting goods stores as an aerosol under the name Permanone® Repel or something similar. It is nearly odorless and resistant to degradation by light, heat, or immersion in water. Interestingly, it can maintain its potency for more than 2 weeks, even through

**Fig. A.3.7** Permanone (permethrin) mosquito and tick repellent. (Photo by Jerome Goddard, Ph.D.)

several launderings [14]. Permethrin can be applied to clothing, tent walls, and mosquito nets; in fact, sleeping under permethrin-impregnated mosquito nets has been tried extensively in malaria prevention campaigns in Africa, New Guinea, Pakistan, and Malaysia. For personal protection, the combination of permethrin-treated clothing and DEET-treated skin creates almost complete protection against mosquito bites. In field trials conducted in Alaska, persons wearing permethrin-treated uniforms and 35% DEET (on exposed skin) had more than 99.9% protection (1 bite/hour) over 8 hours, whereas unprotected persons received an average of 1188 bites/hour [15]. Permethrin is extremely effective against ticks. In one study, a pressurized spray of 0.5% permethrin was compared to 20 and 30% DEET products on military uniforms in a highly infested tick area. A 1-minute application of permethrin provided 100% protection, compared to 86 and 92% protection with the two DEET products, respectively [16].

# References

1. Reimer LJ, Thomsen EK, Tisch DJ, Henry-Halldin CN, Zimmerman PA, Baea ME, et al. Insecticidal bed nets and filariasis transmission in Papua New Guinea. N Engl J Med. 2013;369(8):745–53.
2. N'Guessan R, Corbet V, Akogbeto M, Rowland ME. Reduced efficacy of insecticide-treated nets and indoor residual spraying for malaria control in pyrethroid resistance area, Benin. Emerg Infect Dis. 2007;13:199–206.
3. Fradin MS. Mosquitoes and mosquito repellents: A clinician's guide. Ann Intern Med. 1998;128:931–40.
4. Chou JT, Rossignol PA, Ayres JW. Evaluation of commercial insect repellents on human skin against Aedes aegypti. J Med Entomol. 1997;34:624–30.
5. Buescher MD, Rutledge LC, Wirtz RA, Nelson JH. The dose-persistence relationship of DEET against Aedes aegypti. Mosq News. 1983;43:364–6.
6. Anonymous. Insect repellents. Med Lett Drugs Ther. 2016;58:83–4.
7. Evans SR, Korch GW Jr, Lawson MA. Comparative field evaluation of permethrin and DEET-treated military uniforms for personal protection against ticks. J Med Entomol. 1990;27:829–34.
8. Evans SR. Personal protective techniques against insects and other arthropods of military significance. U.S. Army Environmental Hygiene Agency, Aberdeen Proving Ground, Maryland, TG No. 174, p. 90; 1991.
9. CDC. Updated Information Regarding Mosquito Repellents. U.S. Centers for Disease Control and Prevention, Atlanta, GA (http://www.cdc.gov/ncidod/dvbid/westnile/repellentupdates.htm); 2009.
10. Frances SP, Waterson DG, Beebe NW, Cooper RD. Field evaluation of repellent formulations containing deet and picaridin against mosquitoes in northern territory, Australia. J Med Entomol. 2004;41(3):414–7.
11. Klun JA, Khrimian A, Margaryan A, Kramer M, Debboun M. Synthesis and repellent efficacy of a new chiral piperidine analog: comparison with Deet and Bayrepel activity in human-volunteer laboratory assays against *Aedes aegypti* and *Anopheles stephensi*. J Med Entomol. 2003;40(3):293–9.
12. Barnard DR, Xue RD. Laboratory evaluation of mosquito repellents against Aedes albopictus, Culex nigripalpus, and Ochierotatus triseriatus (Diptera: Culicidae). J Med Entomol. 2004;41(4):726–30.

13. Fradin MS, Day JF. Comparative efficacy of insect repellents against mosquito bites. N Engl J Med. 2002;347:13–8.
14. Schreck CE, McGovern TP. Repellents and other personal protection strategies against *Aedes albopictus*. J Am Mosq Control Assoc. 1989;5:247–50.
15. Lillie TH, Schreck CE, Rahe AJ. Effectiveness of personal protection against mosquitoes in Alaska. J Med Entomol. 1988;25:475–8.
16. Schreck CE, Snoddy EL, Spielman A. Pressurized sprays of permethrin or deet on military clothing for personal protection against *Ixodes dammini*. J Med Entomol. 1986;23:396–9.

# Index

© Springer International Publishing AG, part of Springer Nature 2018
J. Goddard, *Infectious Diseases and Arthropods*, Infectious Disease,
https://doi.org/10.1007/978-3-319-75874-9